New Boson Quantum Field Theory, Dark Matter Dynamics, Dark Matter Fermion Layer Mixing, Genesis of Higgs Particles, New Layer Higgs Masses, Higgs Coupling Constants, Non-Abelian Higgs Gauge Fields

Physics is Logic VII

Stephen Blaha

Blaha Research

ISBN: 978-0-9970761-3-4

Cover Credits

Rev. 00/00/01 March 20, 2016

To My Parents
Stephen and Wanda Blaha

Some Other Books by Stephen Blaha

All the Megaverse! Starships Exploring the Endless Universes of the Cosmos using the Baryonic Force (Blaha Research, Auburn, NH, 2014)

All the Universe! Faster Than Light Tachyon Quark Starships & Particle Accelerators with the LHC as a Prototype Starship Drive Scientific Edition (Pingree-Hill Publishing, Auburn, NH, 2011).

The Algebra of Thought & Reality: The Mathematical Basis for Plato's Theory of Ideas, and Reality Extended to Include A Priori Observers and Space-Time; Second Edition (Pingree-Hill Publishing, Auburn, NH, 2009)

Universes and Megaverses: From a New Standard Model to a Physical Megaverse; The Big Bang; Our Sister Universe's Wormhole; Origin of the Cosmological Constant, Spatial Asymmetry of the Universe, and its Web of Galaxies; A Baryonic Field between Universes and Particles; Flatverse Extended Wheeler-DeWitt Equation (Blaha Research, Auburn, NH, 2014)

PHYSICS IS LOGIC PAINTED ON THE VOID: Origin of Bare Masses and The Standard Model in Logic, U(4) Origin of the Generations, Normal and Dark Baryonic Forces, Dark Matter, Dark Energy, The Big Bang, Complex General Relativity, A Megaverse of Universe Particles (Blaha Research, Auburn, NH, 2015).

PHYSICS IS LOGIC Part II: The Theory of Everything, The Megaverse Theory of Everything, U(4)⊗U(4) Grand Unified Theory (GUT), Inertial Mass = Gravitational Mass, Unified Extended Standard Model and a New Complex General Relativity with Higgs Particles, Generation Group Higgs Particles (Blaha Research, Auburn, NH, 2015).

The Origin of Higgs ("God") Particles and the Higgs Mechanism: Physics is Logic III, Beyond Higgs – A Revamped Theory With a Local Arrow of Time, The Theory of Everything Enhanced, Why Inertial Frames are Special, Universes of the Mind (Blaha Research, Auburn, NH, 2015).

The Origin of the Eight Coupling Constants of The Theory of Everything: U(8) Grand Unified Theory of Everything (GUTE), S^8 Coupling Constant Symmetry, Space-Time Dependent Coupling Constants, Big Bang Vacuum Coupling Constants, Physics is Logic IV (Blaha Research, Auburn, NH, 2015).

New Types of Dark Matter, Big Bang Equipartition, and A New U(4) Symmetry in the Theory of Everything: Equipartition Principle for Fermions, Matter is 83.33% Dark, Penetrating the Veil of the Big Bang, Explicit QFT Quark Confinement and Charmonium, Physics is Logic V

The Periodic Table of the 192 Quarks and Leptons in The Theory of Everything: The U(4) Layer Group, Physics is Logic VI (Blaha Research, Auburn, NH, 2016).

Available on bn.com, Amazon.com, Amazon.co.uk and other international web sites as well as at better bookstores (through Ingram Distributors).

Preface

This book is based on our Extended Standard Model and our Theory of Everything although many parts are relevant independently of our theory. The contents of the book are described in chapter 1. The book describes fermion Layer physics in some detail, and Higgs Mechanism topics both in the conventional way and in our new pseudoquantum formalism – including the Layer group Higgs Mechanism for fermions and gauge fields as well as non-Abelian pseudoquantization and Higgs Mechanism.

CONTENTS

1. Introduction

The appearance of constants in the Standard Model, and our Extended Standard Model and our Theory of Everything, is a source of some puzzlement. The origins of masses and coupling constants are not known. On the other hand the *form* of the Extended Standard Model and the Theory of Everything can be understood from basic principles as we show in Blaha (2015a) and other books.

Guiding principles for the determination of the fundamental constants is lacking. We can mask our lack of knowledge by tracing some constants, such as particle masses, to the Higgs Mechanism. We can even base the determination of coupling constants on our version of the Higgs Mechanism – pseudoquantized Higgs particles. And yet in a sense we are simply trading constants in the theory for equally unknown constants hidden within Higgs potentials – or within a pseudoquantum vacuum built on coherent boson states. We thus are simply pushing the determination of constants to a presumably lower level.

In this book we range over the features of the pseudoquantum Higgs Mechanism. We begin by describing the "new" pseudoquantum scalar boson theory[1] and then describe important applications of the theory in our Theory of Everything.

We discuss the pseudoquantum Higgs Mechanism (which is easily mapped to the conventional Higgs Mechanism) for Generation group contributions to fermion and gauge boson masses, and for the Layer group contributions as well in detail.

In an attempt to understand the origin of Higgs particles we show that Higgs particles may originate in complex ElectroWeak gauge fields which are made real by the extraction of phases. Then gauge fields are real-valued as usually assumed.

In the case of Strong Interaction gauge fields we find that we cannot extract Higgs particles because Strong Interaction gauge fields are inherently complex since they are functions of complex coordinates in our Extended Standard Model. Thus Strong gauge fields are necessarily massless – an obvious, but generally unremarked, property which we can now explain.

Our pseudoquantum approach is also shown to imply a local Arrow of Time, the special role of inertial reference frames, and inertial mass equals gravitational mass (formerly an assumed principle – now a result of complex General Relativity when decomposed to real General Relativity and Higgs fields).

We describe the full set of fermion Higgs mass contributions: from the ElectroWeak sector, from complex General Relativity, from the Generation group (which also yields the four generations of fermions), and from the Layer group which yields four layers of fermions. Higgs contributions to gauge fields' masses are also described in detail.

---

[1] Pseudoquantum theory is described in S. Blaha, Phys. Rev. **D17**, 994 (1978) and references therein. See Appendix 2-A. It is thus new only in the sense that it has not been used in the past 38 years to the author's knowledge.

The Periodic Table of Fermions is described, and its interactions and Higgs masses are found. As noted in our earlier books it gives the proportion of Dark Matter in the universe to be 83.33% - a result in close agreement with astrophysical estimates. Recent studies of the proportion of Dark matter in the universe have yielded two estimates: 84.5% by Aghanim et al in Astronomy and Astrophysics 1303;5062 and 81.5% from a NASA fit to various models.

Since there is continuing speculation about mass values changing in time (and perhaps in locale) we show our pseudoquantized Higgs Mechanism supports space-time dependent masses should they be found.

We also describe our Higgs Mechanism formalism for the eight coupling constants. Again the mechanism can support constant coupling constants as well as space-time dependent coupling constants (a continuing speculation). We show that the coherent field formulation of the pseudoquantum vacuum states enables us to see that all known coupling constants (including gravity) emerge from constants that are all about one in value!

Layer group mass mixing is described in detail as it introduces the somewhat unfamiliar concept of mass mixing between generations in the four layers. Similarly, the unusual pattern of Layer group interactions is described using a new triplet notation for fermion species s, layer l, and generation g: (s, l, g) that uniquely specifies each of the 192 fermions in our Theory of Everything.

Introducing an affine connection, which can be taken to be an independent variable, we define a non-Abelian pseudoquantum theory that supports non-zero gauge field vacuum expectation values – non-Abelian – Higgs particles and "almost" classical non-Abelian gauge fields. These almost classical gauge fields may be relevant in the study of the quark-gluon plasma that has been created at Brookhaven and CERN recently.

The book ends with a description of a pseudoquantum spin 1 boson theory which can support the Higgs Mechanism.

2. Scalar Boson Pseudoquantum Field Theory

Boson second quantization has the initial problem of the absence of a barrier to the decay of positive energy states to negative energy states since the Pauli Exclusion Principle does not apply to bosons. This problem has been masked ("overcome") by a clever choice of boundary conditions that are embodied in the creation/annihilation momentum space operator conditions:

$$a|0> = 0 \qquad \text{Conventional Approach} \qquad (2.1)$$
$$a^{\dagger}|0> \neq 0$$

In this conventional approach the creation of negative energy boson states is eliminated *ab initio* by these conditions. Yet boson quantum fields still have a conceptual physical cloud hanging over them that spin ½ fields do not. A spin ½ particle cannot transition to negative energy because there is a filled sea of negative energy particles. No additional particles can fall into the sea due to the Pauli Exclusion Principle that forbids two fermions with the same 4-momentum and quantum numbers.

In the case of scalar particles the Pauli Exclusion Principle does not apply and so, *physically*, a *filled* negative energy sea of bosons is not possible and positive energy bosons should be able to transition to negative energy states. This problem was "resolved" by the above definition of boson vacuums to exclude transitions to negative energy. But the rationale for the definition is lacking. Dirac was once asked about this issue many years ago. He said he had a solution to the problem. However he did not present it – presumably in keeping with his well-known taciturn nature. So the issue remained an open question.

In this book and earlier work[2] we showed that a more physically satisfactory method exists for avoiding the negative energy state problem. This method relies on the use of a larger Fock space in which *negative energy states (or partially negative energy states) are interpreted as states containing classical fields or a mix of classical fields and individual boson particles.* This approach resolves the negative energy boson issue and provides a common framework for boson particles and classical boson fields.

2.1 Benefits of the Pseudoquantization Method

One consequence of the pseudoquantization method is that it enables the appearance of a vacuum expectation value for Higgs particles (a constant classical field) to be understood within a truly quantum framework. Another major consequence of this approach is the appearance of a *local* Arrow of Time due to the Higgs mass generation mechanism (chapter 3) – a concept that has been a subject of interest for over one hundred years. A macroscopic arrow of

---

[2] See Appendix 2-A and references therein.

time is often described as a statistical result. But our approach yields an arrow of time at the local single particle level.

The conventional approach to boson field quantization sweeps these issues "under the rug" rather than seeking a deeper justification. It differs from Dirac's implied notion that the issue merited attention. We will discuss the "arrow of time" within the framework of our pseudoquantum method later.

Another important consequence of the pseudoquantization method is that it singles out inertial reference frames when applied to the case of Higgs particles.

Yet another more subtle consequence of boson pseudoquantization is that it provides a rationale/explanation for the presence of ElectroWeak Higgs bosons, *and for their absence for the strong (gluon) interactions. The question of why there are no strong interaction Higgs bosons has not been previously considered to the best of this author's knowledge.*

2.2 Pseudoquantization of Bosons

We will now consider the pseudoquantization of a scalar boson field. The cases of the pseudoquantization of vector and spin 2 (graviton) fields with internal symmetry indices are similar but differ in details. We will consider the vector field case later. In particular, we will consider Higgs vector bosons with non-zero vacuum expectation values – a case that is particularly well handled by our pseudoquantum formalism.

We begin by defining two fields that correspond to a scalar particle: $\varphi_1(x)$ and $\varphi_2(x)$.[3] These fields will be assumed to have the equal time commutators

$$[\varphi_i(x), \pi_j(y)] = i(1 - \delta_{ij})\delta^3(\mathbf{x} - \mathbf{y}) \qquad (2.1)$$
$$[\varphi_i(x), \varphi_j(y)] = 0$$
$$[\pi_i(x), \pi_j(y)] = 0$$

where δ_{ij} is the Kronecker δ and where $\pi_i(x)$ is the canonically conjugate momentum to $\varphi_i(x)$. The fields $\varphi_1(x)$ and $\pi_1(y)$ will be observable classical fields as shown by eqs. 69 and 70 in Appendix 2-A. Appendix 2-A provides a more detailed and comprehensive discussion of topics in this chapter. The fields $\varphi_2(x)$ and $\pi_2(y)$ will not be observables. Thus $\varphi_1(x)$ and $\pi_1(y)$ can both be sharp on the set of physical states.

The lagrangian density for a generic massless, scalar Klein-Gordon particle is:

$$\mathcal{L} = \partial\varphi_1/\partial x_\mu \partial\varphi_2/\partial x^\mu \qquad (2.2a)$$

with hamiltonian density

$$\mathcal{H} = \pi_1\ \pi_2 + \partial\varphi_1/\partial x_i \partial\varphi_2/\partial x^i \qquad (2.2b)$$

---

[3] The subscripts on the fields are not gauge symmetry indices but simply identifiers distinguishing the fields from each other. When we consider non-abelian gauge theories we will present a justification for having two fields associated with each type of boson.

where i labels spatial coordinates, and $\pi_1 = \partial\varphi_2/\partial t$ and $\pi_2 = \partial\varphi_1/\partial t$. Eqs. 2.2 are without a potential or mass term.

The lagrangian and hamiltonian for a massive scalar boson are

$$\mathcal{L} = \partial\varphi_1/\partial x_\mu \partial\varphi_2/\partial x^\mu - m^2 \varphi_1\varphi_2 \quad (2.2c)$$

with hamiltonian density

$$\mathcal{H} = \pi_1 \pi_2 + \partial\varphi_1/\partial x_i \partial\varphi_2/\partial x^i + m^2 \varphi_1\varphi_2 \quad (2.2d)$$

The fields can be fourier expanded in terms of creation and annihilation operators:

$$\varphi_i(\mathbf{x}, t) = \int d^3k \, [a_i(k)f_k(x) + a_i^\dagger(k)f_k^*(x)] \quad (2.3)$$

for i = 1, 2 where

$$f_k(x) = e^{-ik\bullet x}/(2\omega_k(2\pi)^3)^{1/2}$$

with $\omega_k = |\mathbf{k}|$ in the massless case and $\omega_k = (|\mathbf{k}|^2 + m^2)^{1/2\,i}$ for a massive boson.

The creation and annihilation operators satisfy the commutation relations:

$$[a_i(k), a_j^\dagger(k')] = (1 - \delta_{ij})\delta^3(\mathbf{k} - \mathbf{k'}) \quad (2.4)$$
$$[a_i(k), a_j(k')] = 0$$
$$[a_i^\dagger(k), a_j^\dagger(k')] = 0$$

for i, j = 1, 2. The vacuum state |0> satisfies

$$a_1(k)|0\rangle = a_1^\dagger(k)|0\rangle = 0 \quad (2.5)$$
$$a_2(k)|0\rangle \neq 0 \qquad a_2^\dagger(k)|0\rangle \neq 0 \quad (2.6)$$

The dual vacuum state satisfies

$$\langle 0|a_2(k) = \langle 0|a_2^\dagger(k) = 0 \quad (2.7)$$
$$\langle 0|a_1(k) \neq 0 \qquad \langle 0|a_1^\dagger(k) \neq 0 \quad (2.8)$$

Positive energy single particle *ket* states are defined using $a_2^\dagger(k)$ while negative energy ket states are defined using $a_2(k)$. Positive energy single particle *bra* states are defined using $a_1(k)$ while negative energy bra states are defined using $a_1^\dagger(k)$.

2.3 Classical Field States for Bosons

The defining properties of a classical field state are:

$$\varphi_1(x)|\Phi, \Pi\rangle = \Phi(x)|\Phi, \Pi\rangle \quad (2.9)$$
$$\pi_1(x)|\Phi, \Pi\rangle = \Pi(x)|\Phi, \Pi\rangle$$

where $\Phi(x)$ and $\Pi(x)$ are sharp on the states and where $\varphi_1(x)$ is given by eq. 2.3.

A classical c-number field has the form

$$\Phi(\mathbf{x}, t) = \int d^3k\ [\alpha(k)f_k(x) + \alpha^*(k)f_k^*(x)] \tag{2.10}$$

The corresponding classical state is a coherent state with the form

$$| \Phi, \Pi> = C \exp\{\int d^3k\ [\alpha(k)a_2^\dagger(k) + \alpha^*(k)a_2(k)]\}|0> \tag{2.11}$$

and correspondingly for $\Pi(x)$ where C is a normalization constant.

Additional details on coherent states, which differ from conventional coherent states such as those of Kibble[4] and others, can be found in Appendix 2-A.

2.4 Towards the Higgs Mechanism

With the pseudoquantum coherent state formalism, which gives purely classical fields and also quantum fields through the use of φ_2 and its creation and annihilation operators, we now have the machinery to define a mass mechanism without the introduction of a potential whose origin can only be described as dubious.

For we can define a pseudoquantum coherent state that yields a constant, non-zero vacuum expectation value:

$$\varphi_1(x)|\Phi, \Pi> = \Phi|\Phi, \Pi> \tag{2.12}$$

where Φ is a constant. Evaluating a fermion interaction term we find a mass term emerges[5]

$$\bar{\psi}(\varphi_1 + \varphi_2)\psi \rightarrow \bar{\psi}(\Phi + \varphi_2)\psi \tag{2.13}$$

It can also generate a mass for an interaction with a gauge field of the form

$$A^\mu(\varphi_1 + \varphi_2)^2 A_\mu \rightarrow A^\mu(\Phi + \varphi_2)^2 A_\mu \tag{2.14}$$

It also yields a quantum field theoretic interaction that would result in the production of ElectroWeak particles from these scalar fields. (The production of Higgs particles has recently been found at CERN.)

The present formalism[6] provides a clean way to separate the vacuum expectation value of a scalar particle from its quantum field part in contrast to the conventional Higgs Mechanism where one has to separate a Higgs field into parts manually.

---

[4] T. W. B. Kibble, Jour. Math. Phys. **2**, 212 (1961).

[5] When matrix elements with a "vacuum state" such as eq. 1.10 are taken.

[6] To obtain both the vacuum expectation value and the interaction with the quantum part of the pseudoquantum fields we choose to always specify interactions with fermions and gauge fields using both fields: $\varphi = \varphi_1 + \varphi_2$ as seen above.

Appendix 2-A. PseudoQuantization Theory Paper

This appendix has a reproduction of a paper by the author in 1978 that establishes an extended quantum field theory formalism for lagrangian theories that contain both classical and quantum fields. This paper (S. Blaha, Phys. Rev. **D17**, 994 (1978)) provides a basis for theories in which fields have a non-zero vacuum expectation value that introduces a classical value into an otherwise quantum theory.

PHYSICAL REVIEW D VOLUME 17, NUMBER 4 15 FEBRUARY 1978

Embedding classical fields in quantum field theories

Stephen Blaha*
Physics Department, Syracuse University, Syracuse, New York 13210
(Received 2 August 1976; revised manuscript received 7 November 1977)

We describe a procedure for quantizing a classical field theory which is the field-theoretic analog of Sudarshan's method for embedding a classical-mechanical system in a quantum-mechanical system. The essence of the difference between our quantization procedure and Fock-space quantization lies in the choice of vacuum states. The key to our choice of vacuum is the procedure we outline for constructing Lagrangians which have gradient terms linear in the field variables from classical Lagrangians which have gradient terms which are quadratic in field variables. We apply this procedure to model electrodynamic field theories, Yang-Mills theories, and a vierbein model of gravity. In the case of electrodynamics models we find a formalism with a close similarity to the coherent-soft-photon-state formalism of QED. In addition, photons propagate to $t = +\infty$ via retarded propagators. We also show how to construct a quantum field for action-at-a-distance electrodynamics. In the Yang-Mills case we show that a previously suggested model for quark confinement necessarily has gluons with principal-value propagation which allows the model to be unitary despite the presence of higher-order-derivative field equations. In the vierbein-gravity model we show that our quantization procedure allows us to treat the classical and quantum parts of the metric field in a unified manner. We find a new perturbation scheme for quantum gravity as a result.

I. INTRODUCTION

The relation between classical and quantum systems has been a subject of continuing interest over the years: First, in the original development of quantum mechanics, second, in the study of the classical limit and infrared divergences of quantum-electrodynamic processes,[1,2] and third, in recent attempts to construct strong-interaction models of quark confinement which are for the most part either classical field theory models in search of quantization[3] or quantized gluon models wherein quark confinement is a consequence of infrared behavior.[4,5]

We will describe a new quantization procedure (called pseudoquantization) for field theory which is the analog of Sudarshan's method for embedding a classical-mechanical system in a quantum-mechanical system. It can be used with advantage to either embed a classical field theory in a quantum field theory in such a way as to maintain the classical character of the embedded fields (while studying the interaction between the classical and quantum sectors on essentially the same footing), or to quantize a class of field theories, members of which have been used as models for gravity and as models for the strong interaction with quark confinement.[7-9]

We shall begin (Sec. II) by pseudoquantizing a classical simple harmonic oscillator. This case is of particular importance because of the analogy between the mode amplitudes of a quantum field and the coordinates of a set of simple harmonic oscillators which we will take advantage of in later sections.

In Sec. III we describe the pseudoquantization procedure for field theory. We apply it to electrodynamic models and show that the propagation of photons to $t=+\infty$ is necessarily retarded in this formalism. Further, we display a close analogy between the present formalism and the coherent-soft-photon-state formalism[10] of QED.

In Sec. IV we apply the pseudoquantization procedure to a classical Yang-Mills field. The resulting field theory (with a slight but important modification) has been used as a model for the strong interactions with quark confinement.[7-9] We also apply the pseudoquantization procedure to a vierbein model of gravity and obtain a new perturbation theory for quantum gravity.

In Sec. V we show that principal-value propagators naturally arise in certains sectors of pseudoquantized theories thus verifying an *ad hoc* procedure devised to unitarize a model of quark confinement.[7-9] We also show how to construct a quantum version of action-at-a-distance electrodynamics.

We shall now briefly outline the procedure for embedding a classical-mechanical system in a quantum system.[6] Consider a classical Hamiltonian system with one degree of freedom, and commuting canonical variables, x_1 and p_1, which have the equations of motion

$$\dot{x}_1 = -i[x_1, \hat{H}], \tag{1}$$

$$\dot{p}_1 = -i[p_1, \hat{H}], \tag{2}$$

where defining

$$\hat{H} = -i\left(\frac{\partial H(x_1,p_1)}{\partial p_1}\frac{\partial}{\partial x_1} - \frac{\partial H(x_1,p_1)}{\partial x_1}\frac{\partial}{\partial p_1}\right) \tag{3}$$

allows us to write Hamilton's equations in commutator form. With Sudarshan[6] we define

$$x_2 = i\frac{\partial}{\partial p_1} \tag{4}$$

and

$$p_2 = -i\frac{\partial}{\partial x_1} \tag{5}$$

so that

$$[x_1, x_2] = [p_1, p_2] = 0\,, \tag{6}$$

$$[x_1, p_2] = [x_2, p_1] = i\,, \tag{7}$$

and $\hat{H}$ can now be taken to be the operator

$$\hat{H} = \frac{\partial H(x_1, p_1)}{\partial p_1} p_2 + \frac{\partial H(x_1, p_1)}{\partial x_1} x_2\,. \tag{8}$$

It is now apparent that we can take the above quantities and equations of motion to describe a quantum mechanical system with two degrees of freedom in the "coordinate" representation where the "coordinates" are (x_1, p_1) and the canonical momenta are $\Pi = (p_2, -x_2)$. As we will see below the linearity of $\hat{H}$ in the momenta is crucial for the maintenance of the classical character of x_1 and p_1, and for the observability of the phase-space trajectory. Since we choose to identify the physical observables with the commutative algebra of the coordinate operators, x_1 and p_1, we are led to impose the superselection condition that the momenta, Π, are unobservable. As a result the Hamiltonian and other generators of canonical transformations, which are all linear in the momenta, are also unobservable. However, in each case there is an associated dynamical quantity which is observable.

The required unobservability of the momenta restricts the form of the interaction between a classical-made-quantum system and an inherently quantum system to

$$H_{\text{int}} = \Phi_1 x_2 + \Phi_2 p_2 + X\,, \tag{9}$$

where Φ_1, Φ_2, and X are functions of x_1, p_1, and the quantum system variables. The commutation relations of these functions are also constrained[6] by the superselection rule and the commutativity of the classical variables, x_1 and p_1, and their time derivatives. In the next section we will study the simple harmonic oscillator in order to exemplify the quantum-mechanical case described above and also for direct use in the field-theoretic generalizations of subsequent sections.

II. SIMPLE HARMONIC OSCILLATOR

In this section we discuss the embedding of a classical simple harmonic oscillator in a quantum system. We shall see that the space of states for the indefinite-metric classical-made-quantum system is far larger than the set of states of a classical harmonic oscillator. However, there is a subset of coherent states which may be placed in one-to-one correspondence with the classical harmonic-oscillator states. The classical-made-quantum oscillator is necessarily an indefinite-metric quantum theory for the simple physical reason that the classical bound states cannot have quantized energy levels. Indefinite-metric quantum theories normally have severe problems of physical interpretation. The present work raises the possibility of a partial resolution of some of these problems through a reinterpretation of an indefinite-metric quantum system as a system composed of a classical subsystem interacting with an essentially quantum subsystem of positive metric.

The classical simple harmonic oscillator of frequency ω has the Hamiltonian

$$\mathcal{H} = \frac{1}{2m}(p_1^{\,2} + m^2\omega^2 x_1^{\,2})\,, \tag{10}$$

and the motion is described by

$$x_1 = A\sin(\pi t + \delta)\,, \tag{11}$$

where A and δ are constants. To embed this classical system in a quantum-mechanical system we introduce the variables x_2 and p_2, and, using Eq. (8), obtain the quantum Hamiltonian

$$\hat{H} = \frac{1}{m} p_1 p_2 + m\omega^2 x_1 x_2\,. \tag{12}$$

We eliminate constants by defining (for $i = 1, 2$)

$$x_i = \left(\frac{1}{m\omega}\right)^{1/2} Q_i\,, \tag{13}$$

$$p_i = (m\omega)^{1/2} P_i\,, \tag{14}$$

and

$$\hat{H} = H\omega \tag{15}$$

so that

$$H = P_1 P_2 + Q_1 Q_2\,. \tag{16}$$

The raising and lowering operators are defined by

$$a_j = \frac{1}{\sqrt{2}}(Q_j + iP_j)\,, \tag{17}$$

and

$$a_j^\dagger = \frac{1}{\sqrt{2}}(Q_j - iP_j) \tag{18}$$

for $j = 1, 2$. They have the commutation relations

$$[a_i, a_j] = [a_i^\dagger, a_j^\dagger] = 0\,, \tag{19}$$

$$[a_i, a_j^\dagger] = 1 - \delta_{ij} \tag{20}$$

for $i,j=1,2$. As a result H is seen to have the form

$$H=\tfrac{1}{2}(a_1a_2^\dagger+a_2a_1^\dagger+a_1^\dagger a_2+a_2^\dagger a_1)\,. \tag{21}$$

The number operators are defined by

$$N_1=a_2a_1^\dagger \tag{22}$$

and

$$N_2=a_2^\dagger a_1 \tag{23}$$

and are not Hermitian. However, their sum is Hermitian and we see that

$$H=N_1+N_2\,. \tag{24}$$

The number operators have the following commutation relations with the raising and lowering operators:

$$N_ia_j=a_j(N_i+\delta_{ij}-1) \tag{25}$$

and

$$N_ia_j^\dagger=a_j^\dagger(N_i-\delta_{ij}+1) \tag{26}$$

for $i,j=1,2$.

Up to this point we have maintained a symmetry of the dynamics under the exchange of the subscripts, $1\leftrightarrow 2$. Now we must break that symmetry by choosing a vacuum state which is an eigenstate of Q_1 and P_1 or alternately a_1 and $a_1^\dagger$. The commutativity of Q_1 and P_1 permit this. The observability of Q_1 and P_1 for all time requires it. So we define

$$a_1^\dagger|0\rangle=a_1|0\rangle=0\,. \tag{27}$$

As a result $a_2|0\rangle\neq 0$ and $a_2^\dagger|0\rangle\neq 0$. The eigenstates of the number operators are

$$|n_+,n_-\rangle=(a_2^\dagger)^{n_+}(a_2)^{n_-}|0,0\rangle \tag{28}$$

and satisfy

$$N_1|n_+,n_-\rangle=-n_-|n_+,n_-\rangle\,, \tag{29}$$

$$N_2|n_+,n_-\rangle=n_+|n_+,n_-\rangle\,, \tag{30}$$

so that

$$H|n_+,n_-\rangle=(n_+-n_-)|n_+,n_-\rangle\,. \tag{31}$$

The lack of a lower bound to the energy spectrum is in a sense a problem but a necessary one in that it leads to the possibility of bound states with a continuous energy spectrum—a requirement of a faithful representation of the classical oscillator states. There is a subset of coherent states which can be put in a one-to-one relation with the set of classical oscillator states. The defining property of that subset is that its elements are eigenstates of the operators a_1 and $a_1^\dagger$. If we expand an element of that subset in terms of the number eigenstates

$$|z\rangle=\sum_{n_+,n_-=0}^{\infty}f(z|n_+,n_-)|n_+,n_-\rangle \tag{32}$$

and use

$$a_1^\dagger|n_+,n_-\rangle=-n_-|n_+,n_--1\rangle\,, \tag{33}$$

$$a_1|n_+,n_-\rangle=n_+|n_+-1,n_-\rangle \tag{34}$$

to evaluate the eigenvalue equations

$$a_1|z\rangle=iz^*|z\rangle\,, \tag{35}$$

$$a_1^\dagger|z\rangle=-iz|z\rangle\,, \tag{36}$$

we find

$$f(z|n_+,n_-)=\frac{C(iz^*)^{n_+}(iz)^{n_-}}{n_+!\,n_-!}\,, \tag{37}$$

where C is a constant. As a result

$$|z\rangle=C\exp[i(za_2+z^*a_2^\dagger)]|0,0\rangle\,. \tag{38}$$

We shall call the $|z\rangle$ states coherent states because of their close formal resemblance to the coherent states used in the study of the classical limit of harmonic oscillators, and of quantum electrodynamics[11] (which were eigenstates of the lowering operator but not of the raising operator).

Since $[H,a_1]=-a_1$, and $[H,a_1^\dagger]=a_1^\dagger$, it is clear that the (x_1,p_1) phase-space trajectory is sharp on the set of coherent $|z\rangle$ states. The classical trajectory represented by the state $|z\rangle$ is easily seen to be

$$x_1=\left(\frac{2}{m\omega}\right)^{1/2}R\sin(\omega t+\delta) \tag{39}$$

and

$$p_1=(2m\omega)^{1/2}R\cos(\omega t+\delta)\,, \tag{40}$$

where $z=Re^{i\delta}$. The linearity of H in the "momenta", $\Pi=(p_2,-x_2)$, is crucial for the observability of the phase-space trajectory. In fact, the linearity of all generators of canonical transformations in the momenta is necessary if the canonical transformations are not to take states out of the subset of coherent states.

The superselection rule which follows from the unobservability of the momenta, Π, is best approached by a consideration of the momentum-and coordinate-space representations of the coherent states. In the coordinate-space representation we find that Eqs. (35) and (36) give

$$\left[\left(\frac{m\omega}{2}\right)^{1/2}x_1+i\left(\frac{1}{2m\omega}\right)^{1/2}p_1\right]\langle x_1p_1|z\rangle=iz^*\langle x_1p_1|z\rangle \tag{41}$$

and

$$\left[\left(\frac{m\omega}{2}\right)^{1/2}x_1-i\left(\frac{1}{2m\omega}\right)^{1/2}p_1\right]\langle x_1p_1|z\rangle=-iz\langle x_1p_1|z\rangle\,, \tag{42}$$

so that

$$\langle x_1p_1|z\rangle = \sqrt{2}\,\delta\left(x_1 - \left(\frac{2}{m\omega}\right)^{1/2}\text{Im}z\right)$$
$$\times\,\delta(p_1 - (2m\omega)^{1/2}\text{Re}z). \qquad (43)$$

We have normalized $\langle x_1p_1|z\rangle$ so that

$$\langle z'|z\rangle = \int_{-\infty}^{\infty} dx_1dp_1\langle z'|x_1p_1\rangle\langle x_1p_1|z\rangle$$
$$= \delta(\text{Re}z - \text{Re}z')\delta(\text{Im}z - \text{Im}z'). \qquad (44)$$

In momentum space Eqs. (35) and (36) lead to the differential equations

$$\left[\left(\frac{m\omega}{2}\right)^{1/2} i\frac{d}{dp_2} + \left(\frac{1}{2m\omega}\right)^{1/2}\frac{d}{dx_2}\right]\langle x_2p_2|z\rangle = iz^*\langle x_2p_2|z\rangle \qquad (45)$$

and

$$\left[\left(\frac{m\omega}{2}\right)^{1/2} i\frac{d}{dp_2} - \left(\frac{1}{2m\omega}\right)^{1/2}\frac{d}{dx_2}\right]\langle x_2p_2|z\rangle = -iz\langle x_2p_2|z\rangle. \qquad (46)$$

They are easily integrated to give

$$\langle x_2p_2|z\rangle = \frac{1}{\sqrt{2}\pi}\exp\left[-ip_2\left(\frac{2}{m\omega}\right)^{1/2}\text{Im}z\right.$$
$$\left.+\,ix_2(2m\omega)^{1/2}\text{Re}z\right] \qquad (47)$$

with the normalization condition

$$\langle z'|z\rangle = \int_{-\infty}^{\infty} dx_2dp_2\langle z'|x_2p_2\rangle\langle x_2p_2|z\rangle$$
$$= \delta(\text{Re}z - \text{Re}z')\delta(\text{Im}z - \text{Im}z'). \qquad (48)$$

The transformation function between the two representations is

$$\langle x_1p_1|x_2p_2\rangle = \frac{1}{2\pi}\exp(+ip_2x_1 - ip_1x_2), \qquad (49)$$

so that

$$\langle x_1p_1|z\rangle = \int_{-\infty}^{\infty} dx_2dp_2\langle x_1p_1|x_2p_2\rangle\langle x_2p_2|z\rangle. \qquad (50)$$

Each coherent state, $|z\rangle$, is a superselection sector in itself. There is no measurable dynamical variable $F = F(a_1, a_1^\dagger)$ which connects different states:

$$\langle z'|F(a_1, a_1^\dagger)|z\rangle = F(iz^*, -iz)\delta^2(z - z'). \qquad (51)$$

This reflects the lack of a superposition principle in classical mechanics.

The operator formalism for coherent states is incomplete in that we have not defined an inner product. To remedy this deficiency we define the vacuum dual to $|0,0\rangle$ to satisfy

$$\langle 0,0|a_2 = \langle 0,0|a_2^\dagger = 0 \qquad (52)$$

with $\langle 0,0|0,0\rangle = 1$. The dual state corresponding to the physical state, z, we define to be

$$\langle z| = \langle 0,0|\delta(ia_1 + z^*)\delta(ia_1^\dagger - z)$$
$$\equiv \langle 0,0|\int_{-\infty}^{\infty}\frac{d\alpha d\beta}{(2\pi)^2}\exp[i\alpha(\text{Im}z - 2^{-1/2}Q_1)$$
$$+\,i\beta(\text{Re}z - 2^{-1/2}P_1)] \qquad (53)$$

so that Eqs. (48) and (51) follow if we choose $C = 1$.

Sometimes the dynamical state of a classical system is incompletely known and one only has a set of probabilities that the system is at a particular phase-space point at $t = 0$. If we let $P(z)$ be the probability that the system is at a phase-space point corresponding to z (as defined above), then using the properties

$$P(z) \geq 0, \quad \int d^2z\,P(z) = 1 \qquad (54)$$

one sees that a density operator

$$\rho\delta^2(0) = \int d^2z\,|z\rangle P(z)\langle z| \qquad (55)$$

may be defined which satisfies

$$\text{Tr}\rho = 1 \qquad (56)$$

and

$$\langle z'|\rho|z'\rangle \equiv \lim_{z''\to z'}\langle z''|\rho|z'\rangle = P(z'). \qquad (57)$$

The mean value of an observable $A = A(a_1, a_1^\dagger)$ is given by

$$\langle A\rangle = \text{Tr}\rho A = \int d^2z\,A(iz^*, -iz)P(z), \qquad (58)$$

and one can develop a formalism similar to the density-matrix formalism of quantum mechanics.

We now turn to a closer investigation of the relation of the pseudoquantum mechanics discussed above and true quantum-mechanical systems. We shall be particularly interested in the relation of the coherent states described above and the coherent states of a quantum-mechanical harmonic oscillator—to which they bear such a remarkable resemblance. We shall see that the pseudoquantum oscillator system is equivalent to an indefinite-metric quantum system composed of a harmonic oscillator (thus the connection to the coherent-state quantum oscillator formalism) and an "inverted" oscillator to be described below.

Let us define the following rotated raising and lowering operators in terms of the operators defined in Eqs. (17) and (18):

$$b_1 = a_1\cos\theta + a_2\sin\theta, \qquad (59)$$

$$b_2 = -a_1\sin\theta + a_2\cos\theta. \qquad (60)$$

Their commutation relations are

$$[b_1, b_1^\dagger] = \sin(2\theta), \quad (61)$$

$$[b_2, b_2^\dagger] = -\sin(2\theta), \quad (62)$$

$$[b_2, b_1^\dagger] = [b_1, b_2^\dagger] = \cos(2\theta) \quad (63)$$

with all other commutators equal to zero. The Hamiltonian of Eq. (21) becomes

$$H = \tfrac{1}{2}(\{b_1, b_1^\dagger\} - \{b_2, b_2^\dagger\})\sin(2\theta) + \tfrac{1}{2}(\{a_1, a_2^\dagger\} + \{a_2, a_1^\dagger\})\cos(2\theta), \quad (64)$$

where $\{u, v\} = uv + vu$.

Now θ is an arbitrary angle and it is obvious that choosing $\theta = 0$ gives the commutation relations and Hamiltonian studied above. However, the choice $\theta = \pi/4$ results in a new form for H and the commutation relations, which can be interpreted as a harmonic oscillator (the b_1 and $b_1^\dagger$ sector) and an "inverted" harmonic oscillator (the b_2 and $b_2^\dagger$ sector) where the commutator and b_2 terms in the Hamiltonian have the wrong sign. The commutativity of the oscillator raising and lowering operators with the inverted oscillator raising and lowering operators leads to a simple factorization of the coherent states which lays bare the basic of the close similarity of form for our coherent states and the coherent states of a quantum oscillator[10]:

$$|z\rangle = \frac{1}{\sqrt{2\pi}} \exp\left[\frac{i}{\sqrt{2}}(zb_1 + z^* b_1^\dagger)\right] \times \exp\left[\frac{i}{\sqrt{2}}(zb_2 + z^* b_2^\dagger)\right] |0, 0\rangle, \quad (65)$$

while the coherent state of Ref. 11 has the form

$$|\alpha\rangle = \exp(\alpha b^\dagger - \alpha^* b)|0\rangle, \quad (66)$$

where α is a complex numer and $[b, b^\dagger] = 1$. It should be remembered that our choice of vacuum state such that $a_1|0,0\rangle = a_1^\dagger|0,0\rangle = 0$ obviates a simple direct relationship.

Since we have uncovered an interesting relation between a classical-made-quantum system and a "quantum" system of indefinite metric the possibility of reinterpreting indefinite-metric quantum systems as systems containing classical subsystems naturally arises.

III. EMBEDDING OF CLASSICAL FIELDS

In this section we shall discuss the embedding of a classical field theory in a quantum field theory. We shall study the embedding in detail for a scalar field and then describe the features of a classical-made-quantum electrodynamics which we shall call pseudoquantum electrodynamics for the sake of brevity.

Consider a classical field, $\phi_1(x)$, with canonically conjugate momentum, $\pi_1(x)$, and Hamiltonian equations of motion

$$\frac{d}{dt}\phi_1(x) = \frac{\delta \hat{H}}{\delta \pi_1(x)}, \quad (67)$$

$$\frac{d}{dt}\pi_1(x) = \frac{-\delta H}{\delta \phi_1(x)}, \quad (68)$$

where $\hat{H}$ is the Hamiltonian. We wish to define a "quantum" Hamiltonian, H, which allows us to rewrite Eqs. (67) and (68) in commutator form:

$$\frac{d}{dt}\phi_1(x) = i[H, \phi_1(x)], \quad (69)$$

$$\frac{d}{dt}\pi_1(x) = i[H, \pi_1(x)]. \quad (70)$$

Equations (69) and (70) are satisfied if

$$H = \int d^3x \left[\frac{\delta H}{\delta \pi_1(x)} \frac{1}{i} \frac{\delta}{\delta \phi_1(x)} - \frac{\delta H}{\delta \phi_1(x)} \frac{1}{i} \frac{\delta}{\delta \pi_1(x)}\right]. \quad (71)$$

We now formally define

$$\phi_2(x) = i\frac{\delta}{\delta \pi_1(x)} \quad (72)$$

and

$$\pi_2(x) = -i\frac{\delta}{\delta \phi_1(x)}, \quad (73)$$

so that

$$H = \int d^3x \left[\frac{\delta \hat{H}}{\delta \pi_1(x)}\pi_2(x) + \frac{\delta \hat{H}}{\delta \phi_1(x)}\phi_2(x)\right]. \quad (74)$$

The fields satisfy the equal-time commutation relations

$$[\phi_i(x), \pi_j(y)] = i(1 - \delta_{ij})\delta^3(\vec{x} - \vec{y}), \quad (75)$$

$$[\phi_i(x), \phi_j(y)] = 0, \quad (76)$$

$$[\pi_i(x), \pi_j(y)] = 0, \quad (77)$$

where δ_{ij} is the Kronecker δ.

We note that the linearity of H in ϕ_2 and π_2 is necessary to maintain the classical character of ϕ_1 and π_1. This is best seen by an examination of Eqs. (69) and (70) and the corresponding Hamiltonian equations for ϕ_2 and π_2. (Other generators of canonical transformations are also linear in π_2 and ϕ_2.)

$\phi_2(x)$ and $\pi_2(x)$ will not be observables on the set of physical states, so that $\phi_1(x)$ and $\pi_1(x)$ will both be sharp on the set of physical states and satisfy superselection rules.

If we wish to couple the classical field to a truly quantum system and maintain the classical nature of the field then certain restrictions exist on the form of the total Hamiltonian H_{tot} and on the commutation relations of the various terms occurring in it. First, the coupling must satisfy the requirement that H_{tot} is linear in $\phi_2(x)$ and $\pi_2(x)$. If we denote the quantum fields by ψ and write the general form of the Hamiltonian as

$$H_{tot} = H + H_Q(\psi) + H_{int}, \tag{78}$$

where H is given by Eq. (74), $H_Q(\psi)$ depends only on the quantum fields, ψ, and

$$H_{int} = \int d^3x[\tilde{A}(\phi_1, \pi_1, \psi)\phi_2(x) + \tilde{B}(\phi_1, \pi_1, \psi)\pi_2(x) + \tilde{C}(\phi_1, \pi_1, \psi)], \tag{79}$$

then we can rearrange the Hamiltonian so that

$$H_{tot} = \int d^3x[A(\phi_1, \pi_1, \psi)\phi_2(x) + B(\phi_1, \pi_1, \psi)\pi_2(x) + C(\phi_1, \pi_1, \psi)], \tag{80}$$

where

$$A = \frac{\delta \hat{H}}{\delta\phi_1(x)} + \tilde{A}, \tag{81}$$

$$B = \frac{\delta \hat{H}}{\delta\pi_1(x)} + \tilde{B}, \tag{82}$$

and

$$C = \tilde{C} + \mathcal{H}_Q \tag{83}$$

with $H_Q = \int d^3x\, \mathcal{H}_Q$. An examination of the equations of motion of $\phi_1(x)$, $\pi_1,(x)$, and ψ,

$$\frac{d}{dt}\phi_1 = B(\phi_1, \pi_1, \psi), \tag{84}$$

$$\frac{d}{dt}\pi_1 = A(\phi_1, \pi_1, \psi), \tag{85}$$

$$\frac{d}{dt}\psi = i[H_{tot}, \psi], \tag{86}$$

and the second time derivatives of ϕ_1 and π_1, such as

$$\frac{d^2}{dt^2}\phi_1(x) = i[H, B] = \int d^3y\Big(-A\frac{\delta B}{\delta\pi_1(y)} + B\frac{\delta B}{\delta\phi_1(y)} + i\phi_2(y)[A, B] + i\pi_2(y)[B(y), B(x)] + i[C, B]\Big), \tag{87}$$

leads us to require the equal-time commutation relations

$$[A(x), A(y)] = [A(x), B(y)] = [B(x), B(y)] = 0, \tag{88}$$

where $A(x) = A(\phi_1(x), \pi_1(x), \psi(x))$, etc., so that $\phi_1(x)$ and $\pi_1(x)$ are independent of ϕ_2 and π_2 and hence observable for all time. An examination of higher time derivatives of ϕ_1 and π_1 lead to further restrictions on the equal-time commutation relations of A, B, and C. Examples are

$$[A, [C, B]] = 0, \tag{89}$$

$$[B, [C, B]] = 0, \tag{90}$$

$$[A, [C, [C, B]]]] = 0, \tag{91}$$

etc. A sufficient condition for satisfying all relations of this class consists of having equal-time commutation relations with the form

$$[A, C] = F_1(A, B, \phi_1, \pi_1) \tag{92}$$

and

$$[B, C] = F_2(A, B, \phi_1, \pi_1). \tag{93}$$

Finally, we note that another obvious requirement [cf. Eqs. (84) and (85)] for the observability of ϕ_1 and π_1 is that A and B depend only on an (equal-time) commutative subset of the quantum field variables, ψ.

The above restrictions on the equal-time commutation relations have a direct interpretation in terms of Feynman diagrams for quantum corrections to the classical field behavior. For example, consider the interaction of the classical field sector with a scalar quantum field, ψ, expressed in the interaction

$$H_{int} = g\phi_2(x)\psi^2(x). \tag{94}$$

If $H_Q(\psi)$ is the conventional free Klein-Gordon Hamiltonian, then we find that Eq. (92) is not satisfied so that the Green's function for the classical ϕ_1 field receives quantum corrections from vacuum polarization loops of ψ particles and thus loses its classical character.

We now define a Lagrangian appropriate to our pseudoquantum field theory and then verify the reasonableness of our definition, and the pseudoquantization procedure described above, by studying the equivalent path-integral formulation. The Lagrangian corresponding to the pseudoquantum Hamiltonian, H, is

$$L = \int d^3x(\pi_1\mathring{\phi}_2 + \pi_2\mathring{\phi}_1) - H, \tag{95}$$

where $L = L(\phi_1, \mathring{\phi}_1, \phi_2, \mathring{\phi}_2)$ and

$$\pi_1 = \frac{\delta L}{\delta\mathring{\phi}_2}, \tag{96}$$

$$\pi_2 = \frac{\delta L}{\delta \dot{\phi}_1} . \tag{97}$$

The vacuum-vacuum transition amplitude for the field theory corresponding to the H_{tot} of Eq. (78) will be shown to be

$$W = \int \prod_x d\phi_1(x) d\phi_2(x) d\pi_1(x) d\pi_2(x) d\psi(x) \exp(iS) , \tag{98}$$

where $S = \int dt\, L_{tot}$ up to external source terms. We begin by considering the vacuum-vacuum transition amplitude corresponding to H_Q,

$$W_Q = \int \prod_x d\psi(x) \exp(iS_Q) , \tag{99}$$

where ϕ_1 has the character of an external source.

We can now introduce the classical behavior of the ϕ_1 field through functional δ functions

$$\int \prod_x d\psi(x) d\phi_1(x) d\pi_1(x) \delta(B(\phi_1, \pi_1, \psi) - \dot{\phi}_1) \times \delta(A(\phi_1, \pi_1, \psi) + \dot{\pi}_1) e^{iS_Q} , \tag{100}$$

which can be put in the form

$$\int \prod_x d\phi_1(x) d\pi_1(x) d\phi_2(x) d\pi_2(x) \times \exp\left\{ i \int d^4x [(\dot{\phi}_1 - B)\pi_2 - (\dot{\pi}_1 + A)\phi_2] + iS_Q \right\} . \tag{101}$$

After performing a partial integration on the $\dot{\pi}_1\phi_2$ term and discarding a surface term we see that the definition of L in Eq. (95) is correct and that the vacuum-vacuum transition amplitude is indeed given by Eq. (98).

The restrictions on the commutation relations of the various terms in the H_{tot} [expressed in Eqs. (88)-(93)] translate into the requirement that the "quantum completion"[11] of the ϕ_2 field does not take place, i.e., that all N-point functions of the ϕ_2 field are zero:

$$\frac{\delta^n W}{\delta J_2(x_1) \delta J_2(x_2) \cdots \delta J_2(x_n)} = 0 , \tag{102}$$

where J_2 is an external source coupled to ϕ_2.

We now discuss the embedding of a free classical Klein-Gordon field in a quantum field theory. The Lagrangian density is

$$\mathcal{L} = \frac{\partial \phi_1}{\partial x^\mu} \frac{\partial \phi_2}{\partial x_\mu} - m^2 \phi_1 \phi_2 . \tag{103}$$

from which one obtains the Euler-Lagrange equations (for $i = 1, 2$)

$$(\Box + m^2)\phi_i(x) = 0. \tag{104}$$

The canonical momenta are (note that π_2 is conjugate to ϕ_1, etc.)

$$\Pi_i = \dot{\phi}_i \tag{105}$$

for $i = 1, 2$ with the equal-time commutation relations given by Eqs. (75)-(77). We expand the fields in Fourier integrals:

$$\phi_1(\vec{x}, t) = \int d^3k [a_1(k) f_k(x) + a_1^\dagger f_k^*(x)] \tag{106}$$

and

$$\phi_2(\vec{x}, t) = \int d^3k [a_2(k) f_k(x) + a_2^\dagger(k) f_k^*(x)] , \tag{107}$$

where

$$f_k(x) = (2\pi)^{-3/2} (2\omega_k)^{-1/2} e^{-ik \cdot x} \tag{108}$$

with $\omega_k = (\vec{k}^2 + m^2)^{1/2}$. The Fourier component operators satisfy the commutation relations

$$[a_i(k), a_j^\dagger(k')] = (1 - \delta_{ij}) \delta^3(\vec{k} - \vec{k}') \tag{109}$$

and

$$[a_i(k), a_j(k')] = [a_i^\dagger(k), a_j^\dagger(k')] = 0 \tag{110}$$

for $i, j = 1, 2$.

In terms of the Fourier coefficients

$$H = \int d^3x (\dot{\phi}_1 \dot{\phi}_2 + \vec{\nabla}\phi_1 \cdot \vec{\nabla}\phi_2 + m^2 \phi_1 \phi_2) \tag{111}$$

becomes

$$H = \int d^3k\, \omega_k [\{a_1(k), a_2^\dagger(k)\} + \{a_2(k), a_1^\dagger(k)\}] . \tag{112}$$

The analogy between the mode amplitudes of the fields and the raising and lowering operators of the simple harmonic oscillator has been previously remarked. We can therefore use the considerations of Sec. II to establish the spectrum of physical states. The defining properties of a physical state are that $\phi_1(x)$ and $\pi_1(x)$ are sharp on it for all time:

$$\phi_1(x) |\Phi, \Pi\rangle = \Phi(x) |\Phi, \Pi\rangle \tag{113}$$

and

$$\pi_1(x) |\Phi, \Pi\rangle = \Pi(x) |\Phi, \Pi\rangle , \tag{114}$$

where $\Phi(x)$ and $\Pi(x)$ are c-number functions of x:

$$\Phi(x) = \int d^3k [\alpha(k) f_k(x) + \alpha^*(k) f_k^*(x)] \tag{115}$$

and

$$\Pi(x) = -i \int d^3k\, \omega_k [\alpha(k) f_k(x) - \alpha^*(k) f_k^*(x)] \tag{116}$$

with $\alpha(k)$ a c-number function of k.

As a result we are led to define a set of physical states, $|\alpha\rangle$, which are in one-to-one correspondence with the classical solutions of the Klein-Gordon equation and satisfy

$$a_1(k)\,|\alpha\rangle = \alpha(k)\,|\alpha\rangle\,, \tag{117}$$

$$a_1^\dagger(k)\,|\alpha\rangle = \alpha^*(k)\,|\alpha\rangle. \tag{118}$$

In analogy with the states of the simple harmonic oscillator (Sec. II) we further define

$$|\alpha\rangle = C\exp\left\{\int d^3k'[\alpha(k')a_2^\dagger(k') - \alpha^*(k')a_2(k')]\right\}|0\rangle\,, \tag{119}$$

where the vacuum state, $|0\rangle$, satisfies

$$a_1(k)\,|0\rangle = a_1^\dagger(k)\,|0\rangle = 0. \tag{120}$$

The physical states, $|\alpha\rangle$, lie in a space which is the infinite tensor product of single-mode spaces. While ϕ_1 and π_1 are sharp for all time on the subset of physical states, we see that ϕ_2 and π_2 are not and, in fact, when applied to a physical state map it into an unphysical state. The superselection rules are embodied in

$$\langle\alpha'\,|\,\Theta\,|\,\alpha\rangle = \Theta_\alpha\delta^2(\alpha - \alpha')\,, \tag{121}$$

where Θ is the operator corresponding to any observable, Θ_α is its eigenvalue for the state $|\alpha\rangle$, and $\delta^2(\alpha-\alpha')$ is a functional δ function in the real and imaginary parts of $\alpha - \alpha'$. The functional δ functions have their origin in the definition of the dual set of physical states. We define the dual vacuum state $\langle 0|$ by

$$\langle 0\,|a_2(k) = 0 \tag{122a}$$

and

$$\langle 0\,|a_2^\dagger(k) = 0 \tag{122b}$$

for all k with $\langle 0|0\rangle = 1$. The dual state corresponding to $\alpha(k)$ we define by

$$\langle\alpha\,| = \langle 0\,|\prod_k \delta(\alpha(k) - a_1(k))\delta(\alpha^*(k) - a_1^\dagger(k))$$

$$\equiv \langle 0\,|\,\delta(\alpha - a_1)\delta(\alpha^* - a_1^\dagger)\,, \tag{123}$$

so that

$$\langle\alpha'\,|\,\alpha\rangle = \delta^2(\alpha' - \alpha) \tag{124}$$

if $C = 1$.

We have now established a procedure for embedding a classical field in a quantum field theory. Given a Lagrangian, L, for a classical field theory describing a field $\phi_1(x)$, the Lagrangian density for the pseudoquantum field theory, $\mathcal{L}_{PQ}$ is

$$\mathcal{L}_{PQ}(\phi_1, \dot\phi_1, \phi_2, \dot\phi_2) = \frac{\delta L}{\delta\phi_1(x)}\,\phi_2(x) + \frac{\delta L}{\delta\dot\phi_1(x)}\,\pi_2(x) \tag{125}$$

up to a divergence with

$$\pi_2(x) = \frac{\delta}{\delta\dot\phi_1(x)}\int d^3x\,\mathcal{L}_{PQ}\,. \tag{126}$$

In the case of a classical electromagnetic field interacting with a quantum electron field, one pseudoquantum model, which describes some electromagnetic processes, has the Lagrangian

$$\mathcal{L} = -\tfrac{1}{2}F^1_{\mu\nu}F^2_{\mu\nu} + \bar\psi(i\nabla\!\!\!/ - e A\!\!\!/_1 - m_0)\psi\,, \tag{127}$$

where $A_\mu^1(x)$ is the classical electromagnetic field, ψ is the electron field, $A_\mu^2(x)$ is the unobservable auxiliary field, and $F^i_{\mu\nu} = \partial_\nu A^i_\mu - \partial_\nu A^i_\mu$ for $i = 1, 2$. Although our interpretation of the free electromagnetic part of the Lagrangian, $-\frac{1}{2}F^1_{\mu\nu}F^2_{\mu\nu}$, is new, the actual form of this term appeared some time ago in a generalization of electrodynamics by Mie,[12] and was recently used in an Abelian prototype model for quark confinement.[8] The equations of motion are

$$\partial^\mu F^1_{\mu\nu} = 0\,, \tag{128}$$

$$\partial^\mu F^2_{\mu\nu} + eJ_\nu = 0\,, \tag{129}$$

and

$$(i\nabla\!\!\!/ - eA\!\!\!/^1 - m)\psi = 0. \tag{130}$$

The canonical momentum which is conjugate to A_μ^1 is

$$\Pi_\mu^2 = F^2_{0\mu} \tag{131}$$

and that conjugate to A_μ^2 is

$$\Pi_\mu^1 = F^1_{0\mu}\,. \tag{132}$$

We take A_μ^1 and Π_μ^1 to be classical fields which are observable for all time. A_μ^2 and Π_μ^2 are not observable. Note that $\mathcal{L}$ is invariant under the independent gauge transformations

$$A_\mu^1 \to A_\mu^1 + \partial_\mu\Lambda^1(x) \tag{133}$$

and

$$A_\mu^2 \to A_\mu^2 + \partial_\mu\Lambda^2(x). \tag{134}$$

Since $\Pi_0^1 = \Pi_0^2 = 0$, it is apparent that A_0^1 and A_0^2 are c numbers. If we chose the Coulomb gauge for A_μ^1,

$$\vec\nabla\cdot\vec{A}^1 = 0\,, \tag{135}$$

and for A_μ^2'

$$\vec\nabla\cdot\vec{A}^2 = 0\,, \tag{136}$$

then we can establish the equal-time commutation relations

$$[\Pi_i^a(\vec{x},t), A_j^b(\vec{y},t)] = i(1-\delta_{ab}) \times \int \frac{d^3k}{(2\pi)^3} e^{i\vec{k}\cdot(\vec{x}-\vec{y})}\left(\delta_{ij} - \frac{k_i k_j}{|\vec{k}|^2}\right) = i(1-\delta_{ab})\delta_{ij}^{tr}(\vec{x}-\vec{y}) \tag{137}$$

for $a,b = 1,2$ and $i,j = 1,2,3$.

This pseudoquantum field theory describes the dynamics of quantum electron fields interacting with a free, classical electromagnetic field. A typical perturbation theory matrix element would have the form

$$\langle \mathcal{A}',0|T(\bar{\psi}(x)J^{\mu_1}(x_1)A^1_{\mu_1}(x_1)J^{\mu_2}(x_2)A^1_{\mu_2}(x_2)\cdots J^{\mu_n}(x_n)A^1_{\mu_n}(x_n)\psi(y))|\mathcal{A},0\rangle, \tag{138}$$

where $|\mathcal{A},0\rangle$ is the tensor product of an electron vacuum state and an electromagnetic state corresponding to the classical field $\mathcal{A}_\mu(z)$. Because $A^1_\mu(x)$ is sharp on this state, the matrix element becomes

$$\langle 0|T(\bar{\psi}(x)J^{\mu_1}(x_1)\cdots J^{\mu_n}(x_n)\psi(y))|0\rangle \mathcal{A}_{\mu_1}(x_1)\mathcal{A}_{\mu_2}(x_2)\cdots\mathcal{A}_{\mu_n}(x_n) \tag{139}$$

modulo a functional δ function in $\mathcal{A}'-\mathcal{A}$. Thus this model is equivalent to a quantized electron field interacting with an external electromagnetic field.

Another possibility for a model electrodynamics is realized by letting the interaction term in Eq. (127) above be replaced with

$$L_{\text{int}} = -e\bar{\psi}A_2\psi. \tag{140}$$

Because the equivalent of the equal-time commutation relation, Eq. (92), is not true in this model, the A^1_μ field loses its purely classical character due to quantum corrections. However, this model may be of value for the study of the modification of the A^1_μ field resulting from the emission of many soft photons by a current.

Since vacuum polarization effects modify the electromagnetic field in this case we define in-field eigenstates (in the transverse gauge) by

$$\vec{A}^1_{\text{in}}|\mathcal{A}\rangle_{\text{in}} = \vec{\mathcal{A}}_{\text{in}}|\mathcal{A}\rangle_{\text{in}}, \tag{141}$$

where

$$|\mathcal{A}\rangle_{\text{in}} = \exp\left[\int d^3k \sum_{\lambda=1}^{2}(\alpha(k,\lambda)a_2^\dagger(k,\lambda) - \alpha^*(k,\lambda)a_2(k,\lambda))\right]|0\rangle \tag{142}$$

and

$$\vec{\mathcal{A}}_{\text{in}} = \int d^3k \sum_{\lambda=1}^{2}\vec{\epsilon}(k,\lambda)[\alpha(k,\lambda)f_k(x) + \alpha^*(k,\lambda)f_k^*(x)] \tag{143}$$

with

$$\vec{A}^i_{\text{in}} = \int d^3k \sum_{\lambda=1}^{2}\vec{\epsilon}(k,\lambda)[a_i(k,\lambda)f_k(x) + a_i^\dagger(k,\lambda)f_k^*(x)] \tag{144}$$

for $i=1,2$. The vacuum state is defined by

$$a_1(k,\lambda)|0\rangle = a_1^\dagger(k,\lambda)|0\rangle = 0$$

for all k,λ. The interacting field, $\vec{A}^1$, is apparently not sharp on $|\mathcal{A}\rangle_{\text{in}}$ but is sharp on

$$|\mathcal{A}\rangle = U^{-1}(t,-\infty)|\mathcal{A}\rangle_{\text{in}}, \tag{145}$$

where

$$U(t,-\infty) = T\left(\exp\left[-i\int_{-\infty}^{t} d^4x H_{\text{int}}(A^2_{\text{in}},\psi_{\text{in}})\right]\right) \tag{146}$$

because

$$\vec{A}^1(\vec{x},t) = U^{-1}(t,-\infty)\vec{A}^1_{\text{in}}(\vec{x},t)U(t,-\infty). \tag{147}$$

With these preliminaries completed, the study of physical processes within the framework of these models is now possible, although we shall not pursue it in this report.

Before turning to a discussion of non-Abelian gauge field theories, it is worth noting that the choice of vacuum state we have made necessitates a redefinition of normal-ordering. By normal-ordering a Lagrangian term we shall mean that the observable fields (to which we have consistently appended the superscript or subscript one) are to be placed to the right, and unobservable fields, labeled by two, are to be placed to the left. Thus Wick's theorem (with our definition of normal-ordering) becomes in the case of two fields

$$T(\phi_{1\,\text{in}}(x_1)\phi_{2\,\text{in}}(x_2)) = :\phi_{1\,\text{in}}(x_1)\phi_{2\,\text{in}}(x_2): + \langle 0|T(\phi_{1\,\text{in}}(x_1)\phi_{2\,\text{in}}(x_2))|0\rangle = \phi_{2\,\text{in}}(x_2)\phi_{1\,\text{in}}(x_1) + \theta(x_{10}-x_{20})[\phi_{1\,\text{in}}(x_1),\phi_{2\,\text{in}}(x_2)]. \tag{148}$$

Note that the Green's function

$$G(x_1,x_2) = \langle 0|T(\phi_{1\,\text{in}}(x_1)\phi_{2\,\text{in}}(x_2))|0\rangle \tag{149}$$

is necessarily retarded. From this we can conclude that the models of electrodynamics, which we have considered, naturally embody the observed retarded nature of classical electrodynamics. Another way of stating this result is: If classical electrodynamics is to have a pseudoquantum formulation, its Green's functions are necessarily retarded. The origin of the asymmetry is the definition of the vacuums (which is equivalent to a specification of boundary conditions). Just as in classical electrodynamics retarded propagation is implemented by a choice of boundary conditions which do not require a commitment to any specific cosmological model.

Finally we would like to note that the Lagrangian obtained from adding L_{int} of Eq. (140) to the Lagrangian of Eq. (127) is equivalent to the usual Lagrangian of electrodynamics plus a term describing a massless Abelian gauge field with the wrong sign. (This is seen by defining new fields equal to the sum and difference of A_μ^1 and A_μ^2.) This field theory may be quantized following the procedure we have outlined. A_μ^1 loses its classical character due to quantum corrections.

IV. NON-ABELIAN GAUGE THEORIES

In this section we shall describe the procedure for embedding a classical non-Abelian Yang-Mills field in a quantum field theory. Then we will discuss a vierbein formulation of quantum gravity which could have been interpreted as a pseudoquantum field theory for a classical metric field if it were not for one term in the Lagrangian which makes it a truly quantum field theory. Nevertheless we suggest a new canonical quantization procedure based on our pseudoquantum approach.

Consider a classical Yang-Mills field, $A_\mu^1 = \underline{A}_\mu^1 \cdot \underline{T}$, where the jth component of $\underline{T}$ is a matrix representing a generator of a non-Abelian group G in the defining representation with commutation relations

$$[T_j, T_k] = i t_{jkl} T_l. \tag{150}$$

We can define a pseudoquantum field theory, wherein the classical character of A_μ^1 is maintained, which has the Lagrangian density

$$\begin{aligned}\mathcal{L} = {} & \tfrac{1}{2}\underline{F}_{\mu\nu}^1 \cdot \underline{F}^{2\mu\nu} - \tfrac{1}{2}\underline{F}^{2\mu\nu} \cdot (\partial_\mu \underline{A}_\nu^1 - \partial_\nu \underline{A}_\mu^1 + g\underline{A}_\mu^1 \times \underline{A}_\nu^1) \\ & - \tfrac{1}{2}\underline{F}^{1\mu\nu} \cdot (\partial_\mu \underline{A}_\nu^2 - \partial_\nu \underline{A}_\mu^2 + g\underline{A}_\mu^1 \times \underline{A}_\nu^2 - g\underline{A}_\nu^1 \times \underline{A}_\mu^2) \\ & + \bar{\psi}(i\not{\nabla} + g\not{A}^1 - m)\psi,\end{aligned} \tag{151}$$

where ψ is a fermion field. The theory is invariant under the local gauge transformation, $S \in G$,

$$\psi' = S^{-1}\psi, \tag{152}$$

$$A_\mu^{1'} = S^{-1} A_\mu^1 S + \frac{i}{g} S^{-1} \partial_\mu S, \tag{153}$$

$$F_{\mu\nu}^{1'} = S^{-1} F_{\mu\nu}^1 S, \tag{154}$$

$$A_\mu^{2'} = S^{-1} A_\mu^2 S, \tag{155}$$

$$F_{\mu\nu}^{2'} = S^{-1} F_{\mu\nu}^2 S. \tag{156}$$

Except for one important term this Lagrangian with its attendant gauge invariance properties has been suggested as a possible model for the quark-confining strong interaction.[8] Since the omitted term has a masslike character $\Lambda^2 \underline{A}_\mu^2 \cdot \underline{A}^{2\mu}$, where Λ has the dimensions of a mass, it is clear that the strong-interaction model's ultraviolet behavior approaches that of the present pseudoquantum theory if the same quantization procedure is followed in both cases. We shall discuss this question further in the next section and show that the *ad hoc* procedure followed in Ref. 8 leads to the same result as the quantization procedure developed in this report.

The Euler-Lagrange equations of motion which are obtained from $\mathcal{L}$ in the canonical manner are

$$\underline{F}_{\mu\nu}^1 = \partial_\mu \underline{A}_\nu^1 - \partial_\nu \underline{A}_\mu^1 + g\,\underline{A}_\mu^1 \times \underline{A}_\nu^1, \tag{157}$$

$$\underline{F}_{\mu\nu}^2 = \partial_\mu \underline{A}_\nu^2 - \partial_\nu \underline{A}_\mu^2 + g\underline{A}_\mu^1 \times \underline{A}_\nu^2 - g\underline{A}_\nu^1 \times \underline{A}_\mu^2, \tag{158}$$

$$(\partial_\mu + g\,\underline{A}_\mu^1 \times)\underline{F}^{1\mu\nu} = 0, \tag{159}$$

$$(\partial_\mu + g\underline{A}_\mu^1 \times)\underline{F}^{2\mu\nu} + g A_\mu^2 \times \underline{F}^{1\mu\nu} + g\underline{J}^\nu = 0, \tag{160}$$

$$(i\vec{\nabla} + g\not{A}^1 - m)\psi = 0, \tag{161}$$

with the conservation law

$$(\partial_\nu + g\underline{A}_\nu^1 \times)\underline{J}^\nu = 0. \tag{162}$$

The canonical momentum which is conjugate to $\underline{A}_j^1$ is

$$\underline{\Pi}_j^2 = \underline{F}_{0j}^2 \tag{163}$$

and the canonical momentum conjugate to $\underline{A}_j^2$ is

$$\underline{\Pi}_j^1 = \underline{F}_{0j}^1 \tag{164}$$

for $j = 1, 2, 3$. The canonical momentum corresponding to the fields $\underline{A}_0^i$ is zero for $i = 1, 2$. The existence of equations of constraint among the Euler-Lagrange equations implies that not all field components are independent, so that we must isolate the independent components prior to defining the canonical equal-time commutation relations.

Following Ref. 8 we choose to work in the Coulomb gauge, $\nabla_i \underline{A}_i^1 = 0$, and define the field variables

$$\underline{A}_i^2 = \underline{A}_i^{2T} + \underline{A}_i^{2L}, \tag{165}$$

$$\underline{\Pi}_i^a = \underline{\Pi}_i^{aT} + \underline{\Pi}_i^{aL}, \tag{166}$$

where

$$\nabla_i \cdot \underline{A}_i^{2T} = \nabla_i \cdot \underline{\Pi}_i^{aT} = 0 \tag{167}$$

and $a = 1, 2$. Then the nonzero equal-time commutation relations are

$$[\Pi_{ip}^{aT}(x), A_{jq}^{bT}(y)] = i\delta_{pq}(1 - \delta_{ab})\delta_{ij}^{\text{tr}}(\vec{x} - \vec{y}), \tag{168}$$

where p and q are internal-symmetry indices, $a, b = 1, 2$, and $i, j = 1, 2, 3$.

While the classical character of A^1_μ can be maintained with our choice of $\mathcal{L}$, this theory has features due to its non-Abelian nature which make it less trivial and therefore more interesting than the corresponding Abelian theory discussed in the last section. If we follow a procedure similar to that in the Abelian case [Eq. (127)] and introduce a set of states appropriate to the quadratic part of the Lagrangian, then the cubic and quartic Yang-Mills terms in the interaction part of the Lagrangian will act to transform $A^1_{\text{in}\,\mu}$ eigenstates into eigenstates of the interacting field A^1_μ. This is, of course, necessary for the classical Yang-Mills equations of motion to be satisfied. Our formalism, thus, offers a perturbative method for calculating solutions of the classical Yang-Mills equations. In addition, it gives an interesting interpretation to the short-distance behavior of the quark-confining field theory of Ref. 8. At short distances the gluon field A^1_μ effectively decouples from the quark sector and becomes, in effect, a free field. This type of short-distance behavior is certainly not at odds with the seemingly simple behavior observed in hadron processes at high energy. Therefore, it is possible that pseudoquantum field theory may be relevant to the short-distance behavior of hadron interaction. Certainly, it is interesting that elementary fermions fall into two similar groups: those which appear to be individually observable (leptons) and those which are not individually observable (quarks).

We now turn to a consideration of a vierbein model of gravity which has certain close similarities to the pseudoquantum field theories we have been studying. In Weyl's formulation[13] of the Einstein-Cartan theory of gravity a vierbein field, $l^{\mu a}(x)$, is introduced which is the "square root" of the metric tensor

$$g^{\mu\nu} = \eta_{ab} l^{\mu a} l^{\nu b}, \tag{169}$$

where η_{ab} is the constant metric tensor of special relativity, where Roman indices transform as vectors under the SL(2, C) group of local Lorentz transformations, and where Greek indices transform as vectors under general coordinate transformations. It is useful to introduce the constant Dirac matrices, γ_a and $4S_{ab} = i[\gamma_a, \gamma_b]$. Under an SL(2, C) transformation,

$$S = \exp[iC^{ab}(x)S_{ab}], \tag{170}$$

a spinor, $\psi(x)$, becomes

$$\psi' = S\psi. \tag{171}$$

The local nature of the transformation requires the introduction of a gauge field

$$B^{ab}_\mu = -B^{ba}_\mu \tag{172}$$

which transforms inhomogeneously,

$$B_\mu \to SB_\mu S^{-1} - \frac{i}{g} S\partial_\mu S^{-1}, \tag{173}$$

so that a Lorentz transformation gauge-covariant derivative can be defined

$$\nabla_\mu \psi = (\partial_\mu + igB_\mu)\psi, \tag{174}$$

where $B_\mu = B^{ab}_\mu S_{ab}$ and $g = 12\pi G$ where G is Newton's constant. Under a gauge transformation we have

$$l^\mu = l^{\mu a}\gamma_a \to Sl^\mu S^{-1}, \tag{175}$$

so that the gauge-covariant derivative of l^μ is defined to be

$$\nabla_\nu l^\mu = (\partial_\nu + igB_\nu \times) l^\mu, \tag{176}$$

where $B_\nu \times l^\mu = [B_\nu, l^\mu]$. The commutator

$$igB_{\mu\nu} = [\partial_\mu + igB_\mu, \partial_\nu + igB_\nu] \tag{177}$$

transforms homogeneously under a gauge transformation

$$B_{\mu\nu} \to SB_{\mu\nu}S^{-1}, \tag{178}$$

and as a second-rank tensor under general coordinate transformations. With these field quantities we are able to construct a Lagrangian $\mathcal{L}_{\text{Weyl}}$ which reduces to the Einstein Lagrangian for gravity when no matter is present,[13]

$$\mathcal{L} = \mathcal{L}_{\text{Weyl}} + \mathcal{L}_{\text{matter}}, \tag{179}$$

where

$$\mathcal{L}_{\text{Weyl}} = \frac{i}{8l} \operatorname{Tr} l^\mu l^\nu B_{\mu\nu} \tag{180}$$

and where, for example, we might let

$$l\,\mathcal{L}_{\text{matter}} = \bar{\psi}(il^\mu \nabla_\mu + m)\psi \tag{181}$$

with $l = \det(l^{\mu a})$.

We observe that the terms containing derivatives in $\mathcal{L}_{\text{Weyl}}$ are linear in the field B_μ—a suggestive feature in view of our previous discussion. However, the quadratic term in B_μ eliminates the possibility of regarding $\mathcal{L}_{\text{Weyl}}$ as a pseudoquantum field theory for a classical field $l^{\mu a}$. But, regardless of this consideration, the fact that $l^{\mu a}$ is necessarily classical in part leads us to consider quantizing vierbein gravity in a manner which is based on the pseudoquantization procedure described above. Remembering that a successful perturbation theory requires the perturbation to be around known solutions we introduce a quadratic Lagrangian term via

$$\mathcal{L} = \mathcal{L}_0 + (\mathcal{L} - \mathcal{L}_0) = \mathcal{L}_0 + \mathcal{L}_{\text{int}}, \tag{182}$$

where

$$\mathcal{L}_0 = -\tfrac{1}{4} i \operatorname{Tr}(B'_{\mu a} l^\mu \gamma^a + ig[B_a, B_b]\gamma^a\gamma^b) \tag{183}$$

and

$$B'_{\mu a} = \partial_\mu B_a - \partial_a B_\mu . \tag{184}$$

Our plan is to follow the pseudoquantization procedure for the "free" part of the Lagrangian $\mathcal{L}_0$. Therefore we will (i) choose a particular coordinate system (harmonic coordinates) and a particular gauge, the "Lorentz" gauge, $\partial^\mu B_\mu = 0$, (ii) establish equal-time commutation relations, (iii) define a set of eigenstates of $l^{\mu a}$, and (iv) proceed to calculate quantum corrections in perturbation theory.

The equations of motion for the "free" Lagrangian $\mathcal{L}_0$ are

$$\partial_\mu B^{ab}_b - \partial_b B^{ab}_\mu = 0 \tag{185}$$

and

$$\partial_\mu (l^{\mu a}\eta^{\nu b} - l^{\nu a}\eta^{\mu b}) + 2g\,(\eta^{\nu a}B^{cb}_c - \eta^{\nu b}B^{ca}_c - \eta^{ac}B^{\nu b}_c + \eta^{bc}B^{\nu a}_c) = 0 . \tag{186}$$

We work in the gravitational equivalent of the Lorentz gauge of electrodynamics,

$$\partial^\mu B^{ab}_\mu = 0 , \tag{187}$$

and choose harmonic coordinates

$$\partial_\mu l^{\mu a} = \tfrac{1}{2}\, \partial^a \eta_{\sigma\tau} l^{\sigma\tau} . \tag{188}$$

The Green's function associated with Eq. (185) is

$$G_{\alpha ef,\rho\sigma}(x,y) = -\tfrac{1}{2} \int \frac{d^4k}{k^2}\, e^{-ik\cdot(x-y)} g_{\alpha ef,\rho\sigma}(k) , \tag{189}$$

where

$$g_{\alpha ef,\rho\sigma}(k) = k_e\left(\eta_{\alpha\rho}\eta_{f\sigma} + \eta_{\alpha\sigma}\eta_{f\rho} - \eta_{\alpha f}\eta_{\rho\sigma} - \frac{k_\alpha k_\rho \eta_{f\sigma} + k_\alpha k_\sigma \eta_{f\rho}}{k^2}\right) - k_f\left(\eta_{\alpha\rho}\eta_{e\sigma} + \eta_{\alpha\sigma}\eta_{e\rho} - \eta_{\alpha e}\eta_{\rho\sigma} - \frac{k_\alpha k_\rho \eta_{e\sigma} + k_\alpha k_\sigma \eta_{e\rho}}{k^2}\right). \tag{190}$$

In order to relate the above Green's function to a time-ordered product of the quantum fields it is first necessary to introduce a set of coherent states, $|L\rangle$, which are eigenstates of $l^{\mu a}$:

$$l^{\mu a}(x)\,|L\rangle = L^{\mu a}(x)\,|L\rangle , \tag{191}$$

where $L^{\mu a}(x)$ is a c-number function of x. In particular, we define $|\eta\rangle$ to satisfy

$$l^{\mu a}|\eta\rangle = \eta^{\mu a}|\eta\rangle , \tag{192}$$

where $\eta^{\mu a}$ is the constant Lorentz metric tensor of special relativity. Given a state $|L\rangle$ we define the field

$$l^{\mu a}_L = l^{\mu a} - L^{\mu a} . \tag{193}$$

This field corresponds to the quantum part of $l^{\mu a}$ and when applied to the purely classical state $|L\rangle$ has the eigenvalue zero.

We now make the identification

$$iG_{\alpha ef,\rho\sigma}(x,y) = \langle L\,|\,T(B_{\alpha ef}(x), l_{L\rho\sigma}(y))\,|\,L\rangle . \tag{194}$$

If we desire to calculate quantum corrections to $l_{\rho\sigma} = \eta_{\rho\sigma}$ we choose $|L\rangle = |\eta\rangle$. (It should be noted that $G_{\alpha ef,\rho\sigma}$ is independent of the choice of $|L\rangle$ as we have defined it.) Because $l_{L\rho\sigma}(y)$ is sharp on $|L\rangle$ we find that the right side of Eq. (194) becomes

$$iG_{\alpha ef,\rho\sigma}(x,y) = \theta(y_0 - x_0)[l_{\rho\sigma}(y), B_{\alpha ef}(x)] \tag{195}$$

up to a functional δ function. From the form of $\mathcal{L}_0$ we see that the commutator is not zero. It is fully determined by an equal-time commutation relation of $l_{\rho\sigma}$ and $B_{\alpha ef}$ (which by the way is the only nonzero equal-time commutator if the canonical procedure is followed), the equations of motion, and the requirement that it be zero at spacelike distances. The "retarded" form of $G_{\alpha ef,\rho\sigma}$ fixes the integration contour around poles in Eq. (192). The other nonzero Green's function in the free Lagrangian model specified by $\mathcal{L}_0$ is

$$iH^{\mu\nu,\rho\sigma}(x,y) = \langle L\,|\,T(l^{\mu\nu}_L(x), l^{\rho\sigma}_L(y))\,|\,L\rangle . \tag{196}$$

It is nonzero owing to the presence of the $[B_\mu, B_\nu]$ term in $\mathcal{L}_0$. We shall show in the next section that it is a principal-value propagator rather than a Feynman propagator. In coordinate space this results in $H^{\mu\nu,\rho\sigma}$ being the sum of the advanced and retarded propagators. As a result our model is equivalent to an action-at-a-distance theory in some sectors.

The classical part of $l_{\mu a}$ is the solution of the classical linearized field equations with appropriate matter sources. The linearized field equations are derived from a Lagrangian consisting of $\mathcal{L}_0$ plus matter terms. (Note that the form of $\mathcal{L}_0$ is obtained by substituting $l_{\mu a} = \eta_{\mu a} + h_{\mu a}$ in $\mathcal{L}_{\text{Weyl}}$, expanding, and keeping quadratic terms.) Thus the class of possible background metrics is restricted.

A simplification occurs in perturbation theory when the classical part of $l_{\mu\alpha}$ is $\eta_{\mu\alpha}$. In this case $(\mathcal{L}_{\text{Weyl}} - \mathcal{L}_0)|\eta\rangle = 0$ when $\mathcal{L}_0$ and $\mathcal{L}_{\text{Weyl}}$ are expressed in terms of asymptotic fields.

V. PRINCIPAL-VALUE PROPAGATORS AND ACTION AT A DISTANCE

In this section we shall show that certain propagators, in field theories where the pseudoquantization procedure has been followed, are principal-value propagators (i.e., the sum of the advanced and retarded Green's functions in coordinate space) rather than Feynman propagators. We also describe a quantum field theory for action-at-a-distance electrodynamics which completes the program initiated by Schwarzschild, Tetrode, and Fokker.[14]

To illustrate the origin of the principal-value propagator we return to the scalar field model of Eq. (103) which described a classical field, $\phi_1(x)$. We introduce an interaction term

$$L_{\text{int}} = -\int d^3z\, \tfrac{1}{2}\lambda^2 [\phi_2(z)]^2 \tag{197}$$

(where λ is a constant), which destroys the purely classical nature of ϕ_1. Suppose we consider the Green's function

$$i\bar{G}(x,y) = \langle 0|\, T(\phi_1(x)\phi_1(y))\,|0\rangle , \tag{198}$$

which would be zero if L_{int} were not present. In terms of in-fields we have

$$i\bar{G}(x,y) = \left\langle 0 \left| T\left(\phi_{1\text{in}}(x)\phi_{1\text{in}}(y)\exp\left(i\int dt\, L_{\text{int}}\right)\right)\right| 0\right\rangle , \tag{199}$$

where the vacuum states, $|0\rangle$ and $\langle 0|$, are defined as in Eqs. (120) and (122). From the definition of the vacuum we find (dropping "in" labels)

$$i\bar{G}(x,y) = \frac{-i\lambda^2}{2}\int d^4z \langle 0|\, T(\phi_1(x)\phi_1(y)\phi_2{}^2(z))\,|0\rangle , \tag{200}$$

which becomes

$$i\bar{G}(x,y) = \frac{-i\lambda^2}{2}\,\epsilon(x_0 - y_0)\,\frac{\partial}{\partial m^2}\,\Delta(x-y) \tag{201}$$

with

$$\Delta(x-y) = -i\int \frac{d^4k}{(2\pi)^3}\,\delta(k^2 - m^2)\epsilon(k_0)e^{-ik\cdot(x-y)} . \tag{202}$$

Using

$$\tfrac{1}{2}\epsilon(x_0 - y_0)\Delta(x-y) = \int \frac{d^4k}{(2\pi)^4}\,\mathrm{P}\,\frac{1}{k^2 - m^2} \times e^{-ik\cdot(x-y)} , \tag{203}$$

we see that

$$\bar{G}(x,y) = -\lambda^2 \int \frac{d^4k}{(2\pi)^4}\,\mathrm{P}\,\frac{1}{(k^2-m^2)^2}\,e^{-ik\cdot(x-y)} , \tag{204}$$

where

$$\mathrm{P}\,\frac{1}{(k^2-m^2)^2} = \frac{1}{2}\left[\frac{1}{(k^2-m^2+i\epsilon)^2} + \frac{1}{(k^2-m^2-i\epsilon)^2}\right] . \tag{205}$$

The form of $\bar{G}$ is consistent with the equations of motion:

$$(\Box + m^2)\phi_1 + \lambda^2\phi_2 = 0 , \tag{206}$$

$$(\Box + m^2)\phi_2 = \delta^4(x-y) . \tag{207}$$

The appearance of the principal-value dipole propagator rather than the Feynman dipole propagator in Eq. (204) is useful because it eliminates certain unitarity problems associated with indefinite-metric fields. However, depending on the model under consideration, it could lead to difficulties with causality. To illustrate the manner in which unitarity problems are resolved, consider the interaction of the ϕ_1 dipole field with a scalar quantum field ψ with

$$L'_{\text{int}} = g\phi_1(x)[\psi(x)]^2 . \tag{208}$$

Suppose we consider the subset of in and out states containing arbitrary numbers of ψ particles but no ϕ_1 or ϕ_2 particles. These states have positive metric. If one could systematically exclude indefinite-metric ϕ_1 and ϕ_2 particles from physical states one would avoid negative probabilities and other problems. But the sum over states in a unitarity sum would normally include states with ϕ_1 particles if the ϕ_1 field had Feynman propagators. In the case of principal-value propagators, no intermediate states with ϕ_1 particles occur, since the pole term is not present. The interaction mediated by the ϕ_1 field is a form of action at a distance and ϕ_1 is properly described by the phrase adjunct field, coined by Feynman and Wheeler.[14] A more detailed discussion of the unitarity question is given in Refs. 7 and 8. In those articles a dipole gluon model for quark confinement was proposed which introduced principal-value propagators in an *ad hoc* manner to resolve unitarity problems. It was pointed out that causality problems did not necessarily exist in those models because the non-Abelian dipole gluons were confined for the same reason as the quarks so that—at the worst—there would be unobservable causality violations at distances of the order of hadron dimensions.

The pseudoquantization procedure may be used to construct a quantum field-theoretic version of action-at-a-distance electrodynamics. Consider the Lagrangian

$$\mathcal{L} = -\tfrac{1}{2}F^{\mu\nu}(\partial_\nu A_\mu - \partial_\mu A_\nu) + \tfrac{1}{4}F^{\mu\nu}F_{\mu\nu} + \bar{\psi}(i\not{\partial} - e\not{A} - m_0)\psi . \tag{209}$$

We define the momentum

$$\Pi_\mu = \frac{\delta \mathcal{L}}{\delta \dot{A}^\mu} = F_{0\mu}. \qquad (210)$$

Going to the transverse gauge as in Sec. IV, we define the equal-time commutation relation

$$[\Pi_i(\vec{x},t), A_j(\vec{y},t)] = i\delta_{ij}^{tr}(\vec{x}-\vec{y}). \qquad (211)$$

Suppose we neglect interaction terms in $\mathcal{L}$ for the moment and choose $F_{\mu\nu}$ to be an observable classical field (as it is up to quantum corrections which we neglect) and A_μ to be unobservable (as it is because it is not gauge invariant). Then we follow our pseudoquantization procedure for

$$\mathcal{L}_0 = -\tfrac{1}{2} F^{\mu\nu}(\partial_\nu A_\mu - \partial_\mu A_\nu) + \tfrac{1}{4} F^{\mu\nu}F_{\mu\nu}. \qquad (212)$$

In particular, we define a vacuum such that

$$F_{\mu\nu}|0\rangle = 0, \quad A_\mu|0\rangle \neq 0, \qquad (213)$$

while

$$\langle 0|A_\mu = 0, \quad \langle 0|F_{\mu\nu} \neq 0. \qquad (214)$$

Then

$$iG_{\mu\nu}(x,y) = \langle 0|T(A_\mu(x)A_\nu(y))|0\rangle \qquad (215)$$

would be zero were it not for $F_{\mu\nu}F^{\mu\nu}$ in $\mathcal{L}_0$. In terms of appropriate in-fields it becomes

$$2iG_{\mu\nu}(x,y) = \int d^4z\,(\theta(x_0-y_0)\theta(y_0-z_0) + \theta(y_0-x_0)\theta(x_0-z_0)) \times [A_{\mu\,\text{in}}(x), F_{\alpha\beta\,\text{in}}(z)][A_{\mu\,\text{in}}(y), F_{\text{in}}^{\alpha\beta}(z)]. \qquad (216)$$

Note that we are treating $F_{\mu\nu}F^{\mu\nu}$ in $\mathcal{L}_0$ as an interaction term. The structure of $G_{\mu\nu}(x,y)$ is the same as that of Eq. (200) so we can conclude that

$$G_{\mu\nu}(x,y) = -g_{\mu\nu}\int \frac{d^4k}{(2\pi)^4}\, \mathrm{P}\frac{1}{k^2}\, e^{-ik\cdot(x-y)} \qquad (217)$$

in the Feynman gauge. Thus the action-at-a-distance interaction follows from the pseudoquantization of electrodynamics. The classical character of $F_{\mu\nu}$ is lost owing to quantum corrections resulting from the presence of $J_\mu A^\mu$ in the Lagrangian.

The example we have just studied has a certain parallel in the vierbein model of gravitation studied in the last section. The forms of the Lagrangian and commutation relations are similar. As a result it is clear that

$$D^{\mu\nu,\lambda\sigma}(x,y) \equiv \left\langle L\left|T\left(l_{L\,\text{in}}^{\mu\nu}(x)l_{L\,\text{in}}^{\lambda\sigma}(y)\int d^4z\,\tilde{\mathcal{L}}_{\text{int}}(z)\right)\right|L\right\rangle \qquad (218)$$

with

$$\tilde{\mathcal{L}}_{\text{int}} = \tfrac{1}{4}\, g\,\mathrm{Tr}\,[B_{\mu\,\text{in}}, B_{\nu\,\text{in}}]\gamma^\mu\gamma^\nu \qquad (219)$$

is a principal-value propagator. Therefore we have constructed an action-at-a-distance version of quantum gravity. Our motivation was to take account of the classical part of $l^{\mu a}$ in a way which did not divorce it from the quantum part to which it is intimately related.

VI. CONCLUSION

We have seen that an alternative to Fock-space quantization exists for a class of field theories which have Lagrangian gradient terms which are linear in field variables. A method was also proposed for constructing Lagrangians of that type from classical Lagrangians with gradient terms which are quadratic in field variables. To some extent this process has a parallel in the passage from Klein-Gordon field Lagrangians which are quadratic in derivatives to Dirac field Lagrangians which are linear in derivatives.

The quantization procedure we have outlined is canonical so far as the fields are concerned. We do, however, make a choice of vacuum states which differs from the usual choice. As a result we have found free propagators which were either retarded, or half-advanced and half-retarded. The choice of vacuum state does not in itself preclude the appearance of Feynman propagators. If one has a good reason to modify the canonical commutation relations then it is possible to obtain Feynman propagators.[15] The procedure we have outlined has, therefore, a greater generality than the particular class of models studied in the present work. It can enable one to embed a classical field theory in a quantum field theory in such a way as to maintain its classical character. It can also be applied to study classical field theories which obtain quantum corrections. Finally it can be applied in order to obtain a fully second-quantized field theory (cf. Ref. 15).

ACKNOWLEDGMENT

This work was supported in part by the U.S. Energy Research and Development Administration.

*Present address: Physics Department, Williams College, Williamstown, Mass. 01267.

[1]D. R. Yennie, S. C. Frautschi, and H. Suura, Ann. Phys. (N.Y.) 13, 379 (1961).

[2]R. J. Glauber, Phys. Rev. 131, 2766 (1963).

[3]W. A. Bardeen, M. S. Chanowitz, S. D. Drell, M. Weinstein, and T.-M. Yan, Phys. Rev. D 11, 1094 (1975).

[4]J. M. Cornwall and G. Tiktopoulos, Phys. Rev. D 13, 3370 (1976).
[5]S. Blaha, Phys. Lett. 56B, 373 (1975).
[6]E. C. G. Sudarshan, Center for Particle Theory report Univ. of Texas—Austin, 1976 (unpublished).
[7]S. Blaha, Phys. Rev. D 10, 4268 (1974).
[8]S. Blaha, Phys. Rev. D 11, 2921 (1975).
[9]S. Blaha, Lett. Nuovo Cimento 18, 60 (1977).
[10]Cf. Ref. 2; T. W. B. Kibble, J. Math. Phys. 9, 315 (1968); Phys. Rev. 173, 1527 (1968); 174, 1882 (1968); 175, 1624 (1968);
[11]A. Salam, lecture at Center for Theoretical Studies, Miami, Florida, 1973 (unpublished).
[12]G. Mie, Ann. Phys. (Leipzig) 37, 511 (1912); 39, 1 (1912); 40, 1 (1913); H. Weyl, *Space, Time, Matter* (Dover, N.Y. 1952).
[13]H. Weyl, Z. Phys. 56, 330 (1929); T. W. B. Kibble, J. Math. Phys. 2, 212 (1961); J. Schwinger, Phys. Rev. 130, 1253 (1963); C. J. Isham, A. Salam, and J. Strathdee, Lett. Nuovo Cimento 5, 969 (1972); F. W. Hehl, P. von der Heyde, G. D. Kerlick, and J. Nester, Rev. Mod. Phys. 48, 393 (1976); and references therein.
[14]K. Schwarzschild, Göttinger Nachrichten 128, 132 (1903); H. Tetrode, Z. Phys. 10, 317 (1922); A. D. Fokker, *ibid.* 58, 386 (1929); J. Wheeler and R. P. Feynman, Rev. Mod. Phys. 17, 157 (1945); 21, 425 (1949).
[15]S. Blaha (unpublished).

3. Higgs Mechanism for Particle Masses

3.1 The Enigma of Higgs Particles and the Higgs Mechanism

In our previous work on the Standard Model, and its generalization to The Extended Standard Model described in a series of books entitled *Physics is Logic*, we showed that the fermion spectrum results from Complex Special Relativity, the gauge interactions result from the Reality group, the fermion generations result from the Generation group, the layers of fermions result from the U(4) Layer group, and from the combination with Complex General Relativity in our Theory of Everything. Higgs particles and the Higgs Mechanism were inserted *ad hoc* to generate particle masses and symmetry breaking effects.

The apparent recent discovery of Higgs particles at CERN seems to solidify the existence of the Higgs sector of the Standard Model and of our Extended Standard Model as described in earlier volumes of *Physics is Logic*.[7]

But whence arises Higgs particles? There does not appear to be a more fundamental cause than the need for particle masses obtained through symmetry breaking. And so the Higgs sector was an expedient mechanism. With our method of avoiding divergences in perturbation theory using Two-Tier quantum field theory the need for the Higgs Mechanism appears to have disappeared with the former need for a symmetry breaking mechanism to generate particle masses. The ElectroWeak sector has no divergences in our approach and thus does not need the renormalization program previously developed that was based on symmetry breaking using the Higgs Mechanism.

In chapter 4 we will see that Higgs fields naturally appear for gauge fields that can be made real-valued, and do not appear for one set of gauge fields – the Strong Interaction SU(3) gauge fields that cannot be made real-valued.

In considering the Higgs Mechanism a number of peculiarities appear that diminishes its attractiveness:

1. As remarked above, it is selective in the sense that some gauge fields have associated Higgs particles and utilize the Higgs Mechanism, and some gauge fields do not have associated Higgs particles. In particular, the ElectroWeak gauge fields, the Generation group gauge fields, the Layer group gauge fields, and the complex gravitation fields have associated Higgs particles. The strong interaction (gluon) gauge fields do not. See chapter 4 for the reason.

2. The conventional Higgs potentials have a quadratic mass term of the "wrong" sign plus a quartic interaction term, which together, generate non-zero vacuum expectation values. They obviously accomplish their goal. But the source of these

[7] Blaha (2015a) and (2015b).

potentials, and why they have their form, is unknown. One suspects a fundamental principle should be operative here.

3. One can imagine creating a Higgs microscope at some super-accelerator. Using this microscope in the presence of a (classical) condensate could enable the Uncertainty Principle to be violated. This possibility, in the case of a microscope using electromagnetic fields, was the source of a heuristic argument for the need to quantize the electromagnetic field.[8]

4. The standard formulation of the Higgs Mechanism uses classical fields under the assumption that a path integral formulation justifies their use. While this may be true, the path integral formulation relies on implicit, unstated boundary conditions that obscure the physics of the quantum field theoretic nature of the mechanism. A direct quantum field theoretic study of the Higgs Mechanism is needed and would further elucidate its character. It is possible, and it has been shown in our earlier books, that the apparently "true" mechanism described below reveals a number of important new results in a properly formulated version of the Higgs Mechanism.

3.2. "True" Origin of an Acceptable Mass Creation Mechanism

In this chapter we will describe a new mechanism that utilizes an extension of quantum field theory to include classical fields that we have called *pseudoquantization*[9] and *pseudoquantum field theory.* It combines both quantum and classical fields within the same framework. In this extended theory vacuum expectation values appear as coherent ground states that are strictly classical in nature.

This chapter is based on our 1978 paper that appeared in the peer-reviewed journal *Physical Review D.* The paper is reproduced in Appendix 2-A for the reader's convenience with the kind permission of The American Physical Society.

We suggest the reader skim or read the paper before proceeding, or refer to the preceding chapter for details as necessary while reading this chapter. The paper also presents a new formulation of Quantum Mechanics that incorporates both quantum and classical mechanics within one framework that is of interest in its own right. Recently, experimenters have been investigating the possibility of macroscopic and other strange quantum phenomena. The new formulation is ideally suited for tracing the transition from a quantum to a classical regime. For example, it is applicable to "large n atoms" where the outermost electrons approach classical behavior with an almost continuous energy spectrum.

3.3 Higgs-Like Vacuum Expectation Value Generation of Masses

The Higgs Mechanism is based on the appearance of non-zero, c-number, vacuum expectation values for Higgs fields due to potential terms directly appearing in lagrangians.

---

[8] Heitler (1954) p. 86 provides a good discussion of the need to quantize the electromagnetic field.

[9] This new formalism was first described in S. Blaha, Phys. Rev. D**17**, 994 (1978).

3.3.1 Pseudoquantization of Higgs Particles

We will now consider the pseudoquantization of a scalar particle using two fields in a manner shown in the preceding chapter. It will become a "Higgs" particle with a non-zero vacuum expectation value. (Section III of our paper in Appendix 2-A contains additional detail.)

Using the formalism of the preceding chapter we define $\varphi_1(x)$ and $\varphi_2(x)$[10] for a generic boson suppressing any internal symmetry indices for simplicity. We define a "vacuum state" containing a coherent superposition of the form of eq. 2.11 that satisfies

$$\varphi_1(x)|\Phi, \Pi> = \Phi|\Phi, \Pi> \qquad (2.12)$$

where Φ is a constant. Evaluating a fermion interaction term we find a mass term emerges[11]

$$\bar{\psi}(\varphi_1 + \varphi_2)\psi \rightarrow \bar{\psi}(\Phi + \varphi_2)\psi \qquad (2.13)$$

It can also generate a mass for an interaction with a gauge field of the form

$$A^\mu(\varphi_1 + \varphi_2)^2 A_\mu \rightarrow A^\mu(\Phi + \varphi_2)^2 A_\mu \qquad (2.14)$$

for ElectroWeak and other gauge fields. The φ_2 term leads to the production of Higgs particles in interactions. (The production of Higgs particles that decay into ElectroWeak gauge particles has recently been found at CERN.)

The present formalism provides a clean way to separate the vacuum expectation value of a scalar particle from its quantum field part in contrast to the conventional Higgs Mechanism where one has to separate a Higgs field into parts manually.

To obtain both the vacuum expectation value and the interaction with the quantum part of the pseudoquantum fields we choose to always specify interactions with fermions and gauge fields using $\varphi = \varphi_1 + \varphi_2$ as seen above.

It appears that our formulation of the mass generation mechanism sheds significant light on the reason for the special prominence of inertial frames. Consider massive scalars.[12] Eqs. 2.12 can describe a massive scalar particle. If the scalar is massive, then the rest frame particle "vacuum" coherent state below yields a non-zero expectation value Φ:

$$|\Phi, \Pi> = C\exp\{[(2\pi)^3 m/2]^{1/2}\Phi[a_2^\dagger(\mathbf{0},m) + a_2(\mathbf{0},m)]\}|0> \qquad (3.1)$$

where m is a generic mass. (We note that the conventional Higgs Mechanism also has mass terms.) *Thus our pseudoquantum formalism allows us to define coherent "vacuum" states that lead to particle masses and Higgs particles.*

---

[10] The subscripts on the fields are not gauge symmetry indices but simply identifiers distinguishing the fields from one another.

[11] When matrix elements with a "vacuum state" such as eq. 2.12 are taken.

[12] Experiments at CERN have apparently discovered a Higgs particle with a 125 GeV/c mass.

3.4 Why Inertial Reference Frames are Special

The great physicists of the early 20th century raised numerous questions about Special Relativity after Einstein and Poincarè's discovery. Prominent among them was the question of why inertial reference frames are of especial importance in Special Relativity, and afterwards in General Relativity.

It appears that our formulation of the mass generation mechanism sheds significant light on the reason for the special prominence of inertial frames. Earlier we considered the case of a massless pseudoquantum scalar. We now consider massive scalars since experiments at CERN have apparently discovered a Higgs particle with a 125 GeV/c mass. The above equations describe a massive scalar particle. If the scalar is massive, then a "vacuum" coherent state, which yields a non-zero expectation value, exists for a particle of mass m in its rest frame.

Then, having established a preferred frame for a Higgs particle, in The Extended Standard Model, and requiring that invariant intervals

$$ds^2 = dt^2 - \mathbf{dx}^2 \quad \text{(in rectangular coordinates)} \tag{3.2}$$

are unchanged by a (complex or real) Lorentz transformation, we find that inertial reference frames are singled out as "special" in the sense that they are the only accessible reference frames that can be generated by a Lorentz boost/transformation from a Higgs particle rest frame. *The Higgs particle vacuum state singles out the class of inertial reference frames.*

Thus Higgs particles play a central role in establishing the basis of physical reality.

3.5 T Invariance of Our Pseudoquantum Scalar Particle Theory

Pseudoquantum scalar particle hamiltonian equations are invariant under time reversal: $t \rightarrow t' = -t$. The vacuum states defined by eqs. 2.1 – 2.8 break the time reversal invariance of the scalar pseudoquantum theory resulting in retarded particle propagators.

The hamiltonian equations

$$[H, \varphi_1(\mathbf{x}, t)] = -i\partial\varphi_1/\partial t \tag{3.3}$$
$$[H, \varphi_2(\mathbf{x}, t)] = -i\partial\varphi_2/\partial t$$

are invariant under time reversal. If we define a time reversal operator transformation U then the time reversed equations are

$$[UHU^{-1}, \varphi_1(\mathbf{x}, -t)] = +i\partial\varphi_1(\mathbf{x}, -t)/\partial(-t) \tag{3.4}$$
$$[UHU^{-1}, \varphi_2(\mathbf{x}, -t)] = +i\partial\varphi_2(\mathbf{x}, -t)/\partial(-t)$$

The operator U, which is unitary, transforms H into –H. This operation is legal because the hamiltonian – in this case – is not positive definite and admits negative energy states.[13] Thus

[13] Unlike the usual case of second quantized Klein-Gordon quantum field theory.

$$[H, \varphi_1(\mathbf{x}, -t)] = -i\partial\varphi_1(\mathbf{x}, -t)/\partial(-t) \qquad (3.5)$$
$$[H, \varphi_2(\mathbf{x}, -t)] = -i\partial\varphi_2(\mathbf{x}, -t)/\partial(-t)$$

and the time reversal invariance of the equations of motion is established for this case.

Time reversal invariance is broken by our choice of vacuum states. This choice is necessary to obtain classical field states as we showed in the preceding chapter and in Appendix 2-A. A demonstration of time reversal symmetry breaking is presented in the following chapter where we show the theory has retarded propagators for particle propagation in "in" and "out" asymptotic states.

Within the interaction region the particle propagators are the sum of retarded and advanced parts that combine to yield principal value propagators – not Feynman propagators.

Many years ago Feynman and Wheeler championed principal value propagators for electrodynamics to obtain an action-at-a-distance theory of Quantum Electrodynamics. While their theory and ours differ from the standard quantum field theory approach there is no reason to view them as faulty, or having serious physical defects. The only question is whether nature chooses conventional quantum field theory or pseudoquantum quantum field theory. In our case the need for a classical scalar particle non-zero vacuum expectation values strongly motivates our choice of psedoquantum Higgs particles.

3.6 Retarded Propagators for Our Pseudoquantum Higgs Particles

In the previous chapter we pointed out that our pseudoquantization Higgs theory has an arrow of time due to its boundary conditions as expressed in its definition of the vacuum state and its dual. In this chapter we will show that the theory uses retarded propagators for propagation to and from the interaction region to asymptotic in-states and out-states. Within an interaction region the theory uses half-retarded – half-advanced propagators. The Appendix 2-A paper has a detailed discussion of propagators between eqs. 145 and 149, and also in section V. We will discuss some aspects of the perturbation theory and propagators of our scalar particles in this chapter.

First we note that in-states at $t = -\infty$ are composed of superpositions of $a_2(k)$ and $a_2^\dagger(k)$ creation and annihilation operators since

$$a_2(k)|0\rangle \neq 0 \qquad\qquad a_2^\dagger(k)|0\rangle \neq 0 \qquad (3.6)$$

while out-states are composed of superpositions of $a_1(k)$ and $a_1^\dagger(k)$ creation and annihilation operators:

$$\langle 0|a_1(k) \neq 0 \qquad\qquad \langle 0|a_1^\dagger(k) \neq 0 \qquad (3.7)$$

Consequently when in-state particles (x_1) propagate into the interaction region (x_2) the relevant propagators are retarded propagators with the form

$$G_{in}(x_2, x_1) = \langle 0|T(\varphi_{1\,in}(x_2), \varphi_{2\,in}(x_1))|0\rangle$$
$$= \theta(x_{20} - x_{10})\langle 0|[\varphi_{1\,in}(x_2), \varphi_{2\,in}(x_1)]\,|0\rangle \qquad (3.8)$$

by eq. 148 of Appendix 2-A. Eq. 3.8 is a manifestly retarded propagator. The choice of vacuums clearly results in a time asymmetry giving a retarded propagation reflecting the familiar Arrow of Time.

A similar situation prevails for propagation to out-states (x_3) from the interaction (x_2) region:

$$G_{out}(x_3, x_2) = <0|T(\varphi_{1\ out}(x_3), \varphi_{2\ out}(x_2))|0>$$
$$= \theta(x_{30} - x_{20})<0|[\varphi_{1\ out}(x_3), \varphi_{2\ out}(x_2)]\ |0> \qquad (3.9)$$

Within the interaction region the Higgs particles have principal value propagators as shown in section V of Appendix 2-A,

Thus we find pseudoquantum Higgs particles embody a local Arrow of Time. The locality of the Arrow of Time is embodied in all the particles that interact with the Higgs particle. Since the mass of *every* particle – bosons and fermions – has a Higgs contribution, and thus *every* particle[14] interacts with Higgs particles, the local Arrow of Time permeates The Extended Standard Model as well as the more familiar Standard Model known from experiment. The local Arrow of Time differs from the *macroscopic* Arrow of Time, which is statistical in nature.

3.7 The *Local* Arrow of Time

In the *Physics is Logic* monographs we saw that complex coordinates led to the form of the fermion spectrum, that the mapping of complex coordinates to real-valued coordinates yielded the Reality group and The Extended Standard Model gauge interaction, that Complex General Relativity led to Higgs particles that were directly united with elementary article masses and gave us the equality of inertial mass and gravitational mass, that the Layer group leads to four layers of fermions, and that the reduction of complex gauge fields to real gauge fields explained the appearance of Higgs fields in The Standard Model and The Extended Standard Model (chapter 4).

Now we see that the pseudoquantization procedure leads to retarded Higgs field propagators and thence to a *local* arrow of time. Many arguments have been put forward over the past hundred plus years for the Arrow of Time. Many arguments based on Statistical Mechanics, Entropy, and Boltzmann's statistical atomic theory have suggested the Arrow of Time is a global statistical consequence. This view seems to contradict the results of elementary particle experiments where a *local* Arrow of Time is evident. It appears that the Arrow of Time has two sources: local and global.

Our rationale for the Arrow of Time begins with retarded Higgs fields. Then we note that Higgs field quantum interactions appear for all fermions and gauge particles. Thus all particle interactions are imbued with an Arrow of Time. Particles that are united to form macroscopic matter inherit their combined local Arrows of Time producing a global Arrow of Time that we experience.

---

[14] Excepting photons.

Thus our pseudoquantiztion approach offers a more satisfactory solution of the origin of the Arrow of Time.

It is remarkable that complex quantities – coordinates and fields – through the Higgs phenomena that we have considered, lead to the equality of inertial mass and gravitational mass (see below), and an Arrow of Time. This unity of mass and time phenomena may reflect the deeper fact that we can have no practical Arrow of Time if all particles were massless, for particle dynamics at light speed would then be pointless. This view has been expressed by DeWitt, Unruh, and others who have pointed out that, physically, time is meaningful and measurable only if masses exist; the larger the mass, the more accurate the time measurement in principle.[15]

3.8 Inertial Mass Equals Gravitational Mass

From the days of Newton through Einstein[16] to the present the equality of gravitational mass and inertial mass has been a topic of interest. Mach, who played an important role, in this ongoing discussion, thought distant masses in the universe were the source of the equality. However the origin of the equality, which has been shown experimentally to very high accuracy, remained uncertain until the Physics is Logic series of books where we show the interconnection of the Extended Standard Model and Complex Gravitation via Higgs generated masses unites gravitational and inertial mass.

In Blaha (2015b) we showed that Complex General Relativity could be formulated in a manner similar to the Extended Standard Model in which the Reality group played a part. This formulation leads to scalar particles that can be viewed as Higgs particles since the fields could be shifted by a constant without affecting the kinetic part of their dynamic equations. If a Higgs potential is present then these fields could undergo spontaneous breakdown and then have non-zero vacuum expectation values.

The decision to base gravitational Higgs fields on the Reality group transformations of Complex General Relativity was based on a desire to build a Theory of Everything. The present Complex Gravity theory, that we have developed, directly entwines gravitation and particle physics through the Reality group. The gravitational Higgs equations then become an elegant, compact unifying feature of a Theory of Everything. The known Higgs equations of the Extended Standard Model are now combined with the gravitational Higgs equations, which are a consequence of Complex General Relativity (suitably extended).[17] The gravitational Higgs potential energy-momentum contribution remains to be justified but can be provisionally inserted by hand just as Higgs potentials are inserted in The Extended Standard Model.

---

[15] No mass, no clock; no clock, no physical time. See Blaha (2015a) pp. 368-371 for a discussion including comments by DeWitt and Unruh.

[16] For example, Einstein and Grossman in 1913 stated, "The theory herein described originates in the conviction that the proportionality between the inertial and gravitational mass of a body is an exact law of nature that must be expressed as a foundation principle of theoretical physics."

[17] This approach is further supported by the use of the Higgs mechanism to produce cosmic inflation and justify the expansion of the universe. See Guth and colleagues for discussions of Higgs induced inflation.

Since fermion field masses are now sums of ElectroWeak Higgs contributions, Generation group Higgs contributions, Layer group Higgs contributions, and gravitational Higgs contributions, and since the gravitational Higgs fields appear in all fermion masses, the equality of inertial and gravitational mass is proven. The gravitational Higgs particles' equations depend, in part, on the gravitational field by eq. 5.50 of Blaha (2015b) and so set the mass scale of the gravitational mass. The presence of the gravitational Higgs contributions ***for all fermions****, sets the scale of the inertial mass Higgs field contributions from the other Extended Standard Model Higgs particles.*

Since an expression cannot mix mass scales, the gravitational mass scale must be the same as the inertial mass scale. Inertial Mass equals gravitational mass.

We have established the equality of inertial and gravitational mass at the short distance quantum level. In our view, this explanation is far more satisfying than basing the equality on a combination of large distance phenomena and quantum phenomena. As Einstein and Weyl have pointed out, all fundamental physics phenomena should be based on a local theory. Complex Gravity as we have constructed it, combined with the Extended Standard Model, furnishes a completely local basic Theory of Everything.

3.8.1 Lagrangian Mass Terms From Blaha (2015b) and (2016a)

In chapter 2 of Blaha (2015b) and in chapter 16 of Blaha (2015a) we discussed the fermion mass contributions from the vacuum expectation values of ElectroWeak and Dark ElectroWeak Higgs particles, and from vacuum expectation values of Generation group Higgs particles.

The form of the species, generations, and layers of fermion mass terms is[18]

$$\mathcal{L}^{Higgs}_{FermionMasses} = \Sigma_{k,a,\alpha,\beta}\bar{\psi}_{kaL\alpha\delta}\eta_k m_{EW_{ka\alpha\beta}}\psi_{kaR\beta} + \Sigma_{k,a,\alpha,\beta}\bar{\psi}_{D_{kaL\alpha}}\eta_{Dk} m_{DEW_{ka\alpha\beta}}\psi_{D_{kaR\beta}} + \qquad \text{ElectroWeak}$$

$$+ \Sigma_{k,a,\alpha,\beta}\bar{\psi}_{UkaL\alpha}\,\eta_{Uka} m_{Uka\alpha\beta}\psi_{UkaR\beta} + \qquad \text{Generation}$$

$$+ \Sigma_{k,a,\alpha,\beta}\bar{\psi}_{DUkaL\alpha}\eta_{DUka} m_{DUka\alpha\beta}\psi_{DUkaR\beta} + \qquad \text{Group U}$$

$$+ \Sigma_{k,g,\delta,\gamma}\bar{\psi}_{LkgL\delta}\,\eta_{Lg} m_{Lg\delta\gamma}\psi_{LkgR\gamma} +$$

$$+ \Sigma_{k,g,\delta,\gamma}\bar{\psi}_{DLkgL\delta}\,\eta_{DLg} m_{DLg\delta\gamma}\psi_{DLkgR\gamma} + \qquad \text{Layer Group L}$$

$$+ \Sigma_{k,a}\bar{\psi}_{GkaL}\eta_{Ga} m_{Gka}\psi_{GkaR} + \Sigma_{k,a}\bar{\psi}_{DGkaL}\eta_{DGa} m_{DGka}\psi_{DGkaR} + \qquad \text{Gravitational}$$

$$+ \text{c.c.} \qquad (5.56')$$

where the subscripts EW, D, U, L and G label ElectroWeak origin, D Dark type, U Generation group origin, L Layer group origin, and G Gravitational origin respectively. The fields labeled η

[18] Layer group contributions have been added to the original eq. 5.56 in Blaha (2015b) in accord with Blaha (2016a).

(with subscripts) are Higgs fields that have non-zero vacuum expectation values.[19] The indices k label species – normal and Dark separately, g labels the (four) generations, and a labels the layers. The index δ and γ label *layer* rows and columns (with implicit sums over generations in the Layer group terms.) The Layer group mass contribution is the same for each fermion in each generation for each species in each layer. The matrices labeled m (with subscripts) are the complex constant mass matrices of species. The indices α, β = 1, … , 4 label *generation* rows and columns.

Eq. 5.56' contains the mass terms for the four layers of fermions in our Theory of Everything. *For each species and generation, the Layer group matrix terms mix the Layer mass contributions.* The three "upper" layers have terms with similar forms but with different mass values. These values are presumably very large. We expect that they are in the TeV and tens of TeVs ranges putting them probably out of range of the current CERN LHC.

Due to the weakness of the ultra-weak interaction, which is mitigated by the anticipated large vacuum expectation values, we expect significant mass cross terms in the mass matrixes of different layers.

The Generation group mass matrices cause mixing between the masses of the four generations in each species. The Layer group mass matrices cause mixing between masses of the four layers of each generation. Fig. 3.1 depicts the mixing of each group's mass terms.

3.8.2 The Full Extended Standard Model Fermion Mass Matrices

Combining the terms in eq. 5.56' for each species we obtain the total mass matrices *for each of the four layers* below. Each mass matrix can then be diagonalized to obtain the masses of the fermions within each of its species.[20]

Charged Lepton Species Total Mass Matrix

$$m_{etot} = m_{EWe} + m_{Ge} + m_{Le}$$

Neutral Lepton Species Mass Matrix

$$m_{\upsilon tot} = m_{EW\upsilon} + m_{G\upsilon} + m_{L\nu}$$

Up-Type Quark Species Mass Matrix (for each color)

$$m_{utot} = m_{EWu} + m_{Uu} + m_{Gu} + m_{Lu}$$

Down-Type Quark Species Mass Matrix (for each color)

$$m_{dtot} = m_{EWd} + m_{Ud} + m_{Gd} + m_{Ld}$$

Dark Charged Lepton Species Total Mass Matrix

$$m_{Detot} = m_{DEWe} + m_{DGe} + m_{DLe}$$

Dark Neutral Lepton Species Mass Matrix

$$m_{D\upsilon tot} = m_{DEW\upsilon} + m_{DG\upsilon} + m_{DL\upsilon}$$

Dark Up-Type Quark Species Mass Matrix

$$m_{Dutot} = m_{DEWu} + m_{DUu} + m_{DGu} + m_{DLu}$$

---

[19] The Higgs fields η… in our pseudoquantum formulation are η… = $\varphi_{1\ldots}(x) + \varphi_{2\ldots}(x)$ as described earlier.

[20] Each Layer's mass contributions are shown.

Dark Down-Type Quark Species Mass Matrix

$$m_{Ddtot} = m_{DEWd} + m_{DUd} + m_{DGd} + m_{DLd}$$

We now note that the preceding formal development yields $m_{Ge} = m_{G\upsilon} = m_{Gu} = m_{Gd} = m_{DGe} = m_{DG\upsilon} = m_{DG\upsilon} = m_{DGu} = m_{DGd} = m_G$. The gravitational mass contribution to all fermions of all species is the same.

Moreover, the gravitational contribution to each fermion mass sets the scale for all fermion masses (and secondarily of massive gauge bosons' masses) yielding the "principle" of Newton, Einstein and others that *inertial mass equals gravitational mass.*

NOTE: The generation group contributions, in the spontaneous breakdown that we described, appear only in quark and Dark quark mass matrices providing, possibly, a reason why quark masses are so much larger than lepton masses.

The mass matrices above can each be diagonalized in a manner similar to that of eqs. 16.50 and 16.51 in chapter 2 of Blaha (2015b) and in Blaha (2015a).

3.9 Space-Time Dependent Particle Masses

It is possible that the Theory of Everything has masses that evolve with time and may also be spatially varying – different values in different parts of the universe. Presently there is no decisive evidence for this possibility although astrophysical studies continue. In this section we will describe the mechanism for space-time dependent masses.

Consider a classical field (time and spatially varying):

$$\Phi(\mathbf{x}, t) = \int d^3k\, [\alpha(k)f_k(x) + \alpha^*(k)f_k^*(x)] \tag{3.10}$$

If we define the coherent vacuum state

$$|\alpha> = C \exp\left\{\int d^3k\, [\alpha(k)a_2^\dagger(k) + \alpha^*(k)a_2(k)]\right\}|0> \tag{3.11}$$

then

$$\varphi_1(x)|\Phi, \Pi> = \Phi(x)|\Phi, \Pi> \tag{3.12}$$
$$\pi_1(x)|\Phi, \Pi> = \Pi(x)|\Phi, \Pi>$$

using the notation of chapter 2 and Appendix 2-A where

$$\varphi_i(\mathbf{x}, t) = \int d^3k\, [a_i(k)f_k(x) + a_i^\dagger(k)f_k^*(x)] \tag{3.13}$$

for i = 1, 2 and

$$f_k(x) = e^{-ik\cdot x}/(2\omega_k(2\pi)^3)^{1/2}$$

with ω_k equal to the energy.

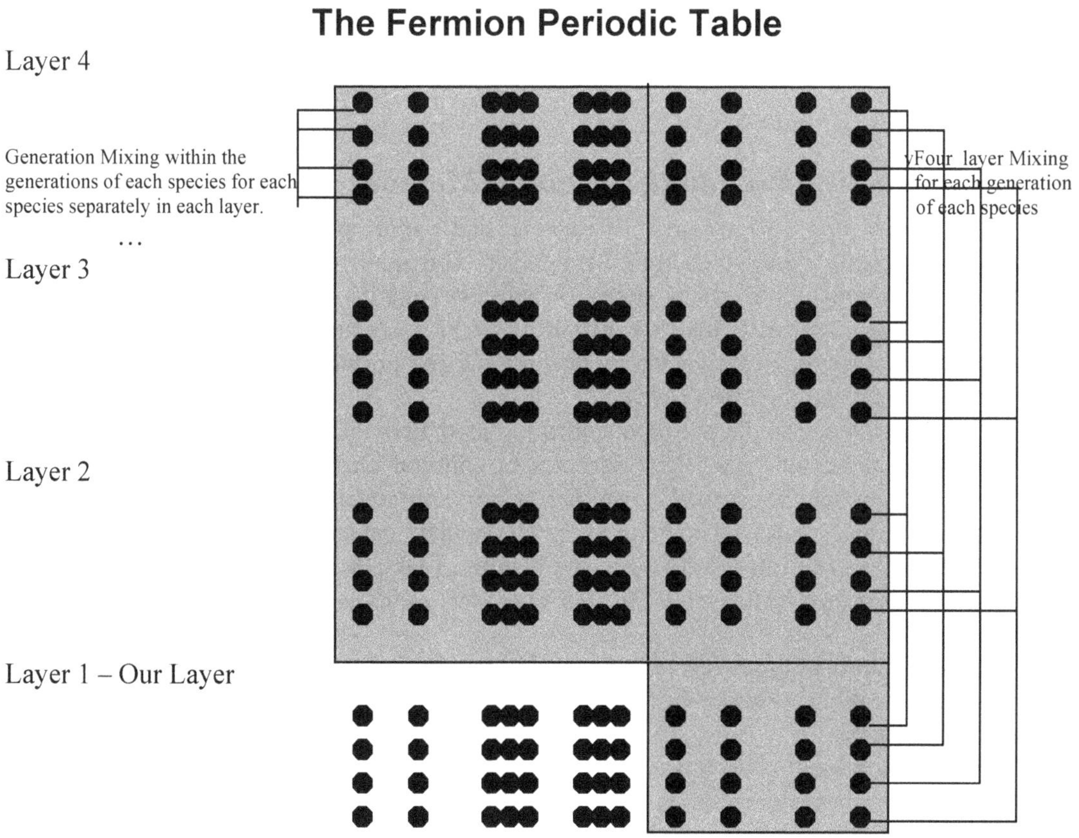

Figure 3.1. Example of pattern of mass mixing of the Generation group and of the Layer group. Dark parts of the periodic table are gray. Light parts are the known fermions with an additional, as yet not found, 4th generation shown. The lines on the left side show an example of the Generation mixing within one species. The Generation mixing applies to each species in each layer. The lines on the right side show an example of Layer mixing within one species with the mixing amongst all four layers of the species for each generation individually.

Eq. 3.11 contains a coherent state $|\alpha>$ for a time and spatially varying mass. The above equations can be generalized to the case of multiple space-time varying masses.[21]

$$| \Phi_1, \Phi_2, \ldots , \Phi_n; \Pi_1, \Pi_2, \ldots , \Pi_n> = C \prod_{i=1}^{n} \exp\left\{\int d^3k \, [\alpha_i(k)a_{2i}^\dagger(k) + \alpha_i^*(k)a_{2i}(k)]\right\}|0> \quad (3.14)$$

Then all n mass vacuum expectation values are space-time dependent:

[21] The "vacuum" state $|0>$ in eq. 3.14 also implicitly has factors for the vacuum expectation values used for fields that give masses to fermions and vector bosons as described in Blaha (2015b).

$$\varphi_{1i}(x) \mid \Phi_1, \Phi_2, \ldots, \Phi_n; \Pi_1, \Pi_2, \ldots, \Pi_n> = \Phi_i(x) \mid \Phi_1, \Phi_2, \ldots, \Phi_n; \Pi_1, \Pi_2, \ldots, \Pi_n> \quad (3.15)$$

Thus our formalism can accommodate space-time varying masses should they be found in the Cosmos.

3.10 Benefits of the Pseudoquantization Method

In this book and earlier work[22] we showed that a more physically satisfactory method for avoiding the negative energy state problem exists. This method relies on the use of a larger Fock space in which negative energy states (or partially negative energy states) are interpreted as states containing classical fields or a mix of classical fields and individual boson particles. This approach resolves the negative energy boson issue and provides a common framework for boson particles and classical boson fields.

One consequence of the pseudoquantization method is that it enables the appearance of a vacuum expectation value for Higgs particles (a constant classical field) to be understood within a truly quantum framework. Another major consequence of this approach is the appearance of a *local* Arrow of Time due to the Higgs mass generation mechanism – a concept that has been a subject of interest for over one hundred years. A macroscopic arrow of time is often described as a statistical result. But our approach yields an arrow of time at the single particle level.

The conventional approach to boson field quantization sweeps these issues "under the rug" rather than seeking a deeper justification. It differs from Dirac's implied notion that the issue merited attention.

Another important consequence of the pseudoquantization method is that it singles out inertial reference frames when applied to the case of Higgs particles.

Yet another more subtle consequence of boson pseudoquantization is that it provides a rationale/explanation for the presence of ElectroWeak Higgs bosons, *and for their absence for the strong (gluon) interactions. The question of why there are no strong interaction Higgs bosons has not been previously considered to the best of this author's knowledge.* See the chapter 5.

[22] See Appendix 2-A and references therein.

Chapter 4. The Broken U(4) Layer Group and Fermion Layers

4.1 The Layer Group

The Generation group was based on U(4) rotations of the four number operators B, L, B_D, and L_D in the one generation Extended Standard Model. We can visualize these rotations horizontally as

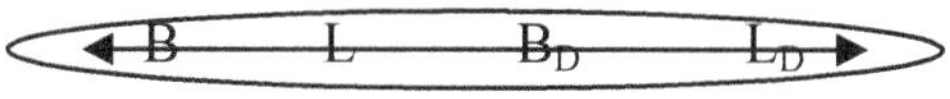

Given U(4) in the one generation case, it is natural to assume that the symmetry applies in a larger case – the case of a four generation Extended Standard Model. We attach a U(4) vector generation index to each fermion field ranging from 1 through 4 and introduce interactions between the generations. The result is the four generation Extended Standard Model presented in Blaha (2015a) and earlier books.

After generations are introduced, then it becomes possible to consider vertical rotations amongst the four generations:

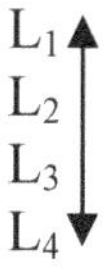

These rotations are based on four conserved numbers L_1, L_2, L_3, and L_4 that count the number of fermions of each generation in a state.[23] L_i counts the number of fundamental fermions in generation i for i = 1, 2, 3, 4. Fermions have positive $L_i = +1$ values and anti-fermions have negative $L_i = -1$ values. For example, if a state has 3 u quarks, 1 d quark, 1 anti-s quark, 2 electrons, 2 anti-τ leptons, 2 Dark muon neutrinos, and one Dark electron neutrino ν_{De} then $L_1 = 3+1+2+1 = 7$, $L_2 = -1+2 = 1$, $L_3 = -2$ and $L_4 = 0$.

It is important to note that the Layer numbers are independent of the baryon and lepton particle numbers that form the basis of the Generation group, and so the physics embodied in the Generation group is not the same as the physics of the Layer group defined below.

Layer numbers are conserved under strong and electromagnetic interactions but broken by the ElectroWeak interactions.

[23] The generations are numbered from 1 to 4 with the lowest masses generation (e, ν_e, u, d) being generation 1.

The partial conservation of the L_i Numbers enables us to define a new broken U(4) symmetry called the Layer group. Thus we have four new (partly) conserved particle numbers. Linear combinations of these numbers are also (partly) conserved:[24]

$$\begin{aligned} L_1' &= aL_1 + bL_2 + cL_3 + dL_4 \\ L_2' &= eL_1 + fL_2 + gL_3 + hL_4 \\ L_3' &= iL_1 + jL_2 + kL_3 + lL_4 \\ L_4' &= mL_1 + nL_2 + oL_3 + pL_4 \end{aligned}$$

using constants labeled from "a" through "p" or

$$L' = AL$$

where A is a 4×4 U(4) matrix.

If we want the new 'primed' set of conserved numbers to be an independent set of numbers then the determinant of the constants must be non-zero. Thus the matrix, A, is invertible.

The set of 4×4 matrices of the above type form a U(4) group.[25] *The choice of U(4) rather than SU(4) is required since there are four independent layer particle numbers and U(4) has four diagonal matrices in its algebra while SU(4) only has three diagonal matrices. U(4) preserves the independence of the four independent particle numbers. Thus we have the Layer group.*

Since the coefficients in Layer group transformations can be local functions, the Layer group is implemented as a Yang-Mills theory.

Just as we extended the reach of the Generation group from one generation to four generations, we can extend the Layer group similarly by assuming that there are four layers of fermions by adding a U(4) vector layer index to each fermion field. Further, since we only know of one layer – our layer – the gauge fields for all vector interactions must be different[26] for each layer. The only exception being gravitation which is universal to all layers. Other layers must then be "Dark" in the sense that each layer is independent except for a new ultra-weak interaction that connects states of different layers.

To implement the Layer symmetry in the Extended Standard Model lagrangian the following steps are required:

---

[24] This discussion parallels the discussion of particle numbers in the following Extract.

[25] If we wish to further limit the values of the 'primed' Numbers to integers assuming the unprimed Numbers are integers then the group of the transformation is the set of permutations of four entities – the Symmetric group S_4. However the 'primed' numbers can be integer or not. There is no apparent physical principle requiring integer Numbers. Quarks are usually assigned Baryon Number 1/3. Also Numbers are not necessarily positive valued.

[26] Later we will see that mass mixing occurs between particles in different layers, which in turn leads to mixing of the gauge fields of the various layers. However the ultra-weakness of the Layer gauge field interactions makes the gauge field mixing negligible – although it is present in principle.

1, All covariant derivatives must acquire another interaction term with 16 U(4) fields corresponding to the 16 generators of the U(4) Layer group.

2. Expand the Theory of Everything to embody the Layer group symmetry by adding another index ranging from 1 through 4 to each fermion field making four layers of four generations of fermions. Thus the fermion fields are in the fundamental representation of the Layer group.

3. Expand the Theory of Everything by adding a Layer group index to each vector gauge field so that each layer has its own complete set of SU(3)⊗SU(2)⊗U(1)⊗SU(2)⊗U(1)⊗U(4) gauge fields. The complex gravitation U(4) Higgs particles are common to all layers. Add Layer indexes to all gauge field dynamic lagrangian terms.

4. Each layer should have its own set of gauge field Higgs particles (modulo miniscule mixing) for other gauge field interactions. One expects that the masses of fermions and gauge fields should be substantially larger for the three 'upper' layers beyond our layer. Otherwise we would have found particles from these upper layers.

5. Insert an interaction term of the form $g_v G_L{}^a V^a{}_\mu$ in the covariant derivative of each fermion using the 16 U(4) Layer gauge fields $V^a{}_\mu$ with a = 1, 2, …, 16 and the 16 U(4) generators $G_L{}^a$ that couple within and between layers using the new layer index where g_v is an ultra-weak coupling constant much smaller than the ElectroWeak coupling constants. This interaction will be between normal matter, between Dark Matter, between normal and Dark matter, and between fermions in the same and different layers. This interaction will be the only interaction between particles in different layers. Its small coupling constant and presumably very large gauge field masses will essentially make the four layers almost independent of each other except for gravitation.

6. Insert Layer group symmetry breaking Higgs fields (independently for all layers) that generate fermion and gauge field mass term contributions independently for each layer.

The above modifications to the Theory of Everything lagrangian will be mathematically similar to the development of the features of the Generation group in Blaha (2015a) presented in the extract below.

The form of the "periodic table" of fermions that results appears in Fig. 3.2 below.

4.2 Layer Group Interactions

The new Layer group interactions play two roles:

1. The four diagonal generator terms of $U_L{}^a V^a{}_\mu$ create transitions amongst the Normal and Dark fermions, including transitions between Normal and Dark, between Normal and Normal, and between Dark and Dark in each layer separately.

2. The non-diagonal generator terms of $U_L{}^a V^a{}_\mu$ create transitions amongst the Normal and Dark fermions between *differing* layers. The non-diagonal interactions are the source of decays from higher layers to lower layers. Since no experimental evidence of such decays has been found as yet we conclude that the interaction constant g_v is ultra-small and/or the Layer gauge boson masses are extraordinarily large. Thus the distribution of fermions in layers may be approximately the same as that at the time of the Big Bang. We considered this issue in Blaha (2016a) and (2016b), and showed the Dark Matter fraction implied by the fermion spectrum (Fig. 3.2) is 83.33% – a result consistent with cosmological data. The fermions of the higher layers must be constituents of "Dark Matter" (as well as the Dark part of our layer) since only the ultra-weak Layer interaction and graviton interactions can connect layers.

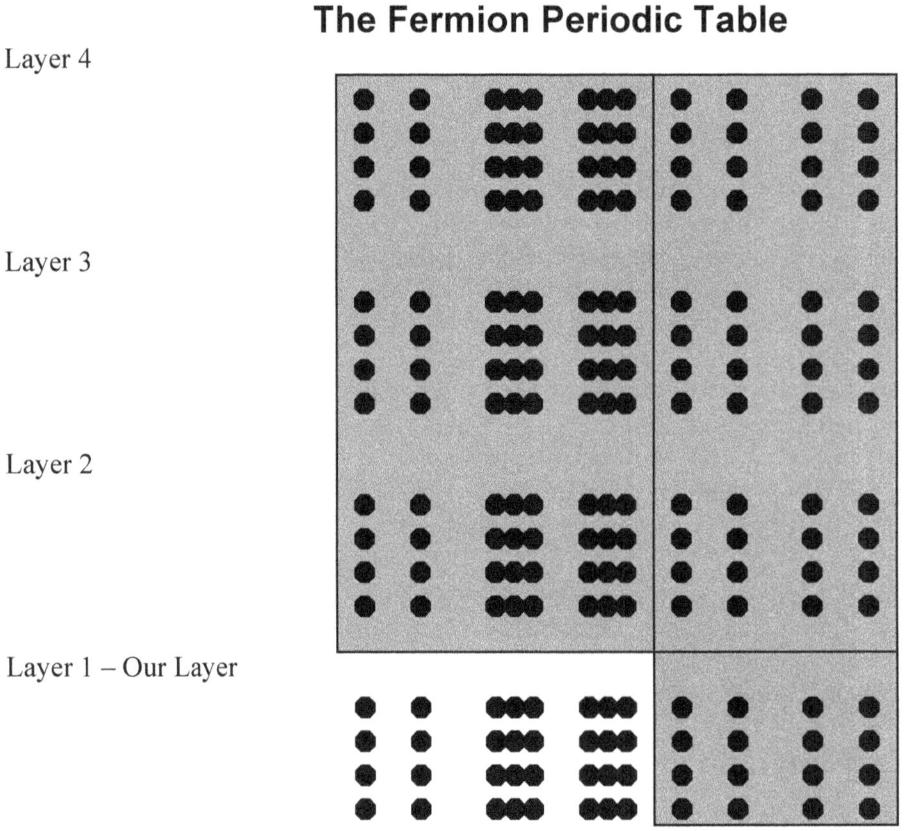

Figure 3.2. Dark parts of the periodic table are gray. Light parts are the known fermions with an additional, as yet not found, 4th generation shown.

Because of the formal similarity of the Generation group analysis to the Layer group analysis that will be presented below we have inserted chapter 16 of Blaha (2015a) with some changes below. All references to equations, chapters, and sections are to those items in Blaha (2015a).

4.3 Extract From Blaha (2015a) With Some Changes

16. Fermion Generations and Broken U(4) Generation Group

In sections 15.2 and 15.3 we showed that a local U(4) symmetry based on conserved fermion particle numbers existed in Nature and added a new assumption of a U(4) Generation group to our construction of the Extended Standard Model. U(4) has 16 generators, which we denote G_i for i = 1, 2, ... , 16. Its fundamental representation has 4×4 matrices. When we introduce this U(4) symmetry directly into the one generation Extended Standard Model each fermion acquires a new index and becomes a four generation set of fermions. Symmetry breaking via a Higgs Mechanism for the U(4) gauge fields gives a different mass to each of the members of each set. The Higgs particle lagrangian for U(4) breaking will be described in section 16.3.

16.1 Four Generation Extended Standard Model

In chapter 14 we derived the form of a one generation Extended Standard Model that included the known parts of the Standard Model (excepting the Higgs sector) and an SU(2)⊗U(1) part for Dark Matter. Dark Matter was linked to normal matter with a simple scalar gauge field. [Now it is replaced by a new U(4) Layer group interaction shown later.]

In this section we generalize to the four generation Extended Standard Model that results.[27] Covariant derivatives acquire another interaction term with 16 U(4) fields U_i^{μ}. In addition we add another index to each fermion field specifying its generation. Lastly a set of initially massless gauge field dynamic terms is added to the Extended Standard Model lagrangian to specify U(4) gauge field evolution.

16.1.1 Two-Tier Lepton Sector

We begin with the definition of a quadruplet of leptons – a pair of doublets, one normal and one Dark, instead of a single doublet. We define left and right lepton quadruplets with[28]

$$\Psi_{L,Ra}(X) = \begin{bmatrix} \psi_{DL,Ra}(X) \\ \psi_{NL,Ra}(X) \end{bmatrix} \qquad (16.1)$$

[27] It is based on the three principles resulting from applying Ockham's Razor ("The simplest choice is often the best."): 1) The only connecting interaction is a weak interaction, 2) The form of ElectroWeak theory remains unchanged, and 3) Dark Matter parallels normal matter in its general characteristics: four generations, SU(3) singlets, an SU(2)⊗U(1) symmetry analogous to ElectroWeak symmetry, SU(2)⊗U(1) Dark lepton and Dark quark doublets.
[28] The X's are Two-Tier coordinates.

where a is a generation index ranging from 1 to 4, where $\psi_{NL,R}(X)$ is a "normal" ElectroWeak-like lepton doublet consisting of a normal electron-like fermion and a normal neutrino-like fermion, and where $\psi_{DL,R}(X)$ is a Dark ElectroWeak-like lepton doublet consisting of a Dark electron-like fermion and a Dark neutrino-like fermion.

We define covariant derivative terms, which we express in matrix form as

$$D_{L,R}(X) = \begin{bmatrix} \gamma^{\mu} D_{DL,R\mu} & 0 \\ 0 & \gamma^{\mu} D_{NL,R\mu} \end{bmatrix} \quad (16.2)$$

where the normal matter left-handed covariant derivative is

$$D_{NL\mu} = \partial/\partial X^{\mu} - \tfrac{1}{2} ig\boldsymbol{\sigma}\cdot\mathbf{W}_{\mu} + \tfrac{1}{2} ig' B_{\mu} - \tfrac{1}{2} ig_G \mathbf{G}\cdot\mathbf{U}_{\mu} \quad (16.3)$$

where g_G is an ultra-weak generation coupling constant, and where $\mathbf{G}\cdot\mathbf{U}_{\mu}$ is the sum of the inner product of 16 U(4) generators G_i and gauge fields $U_{\mu i}(X)$. The Dark matter left-handed covariant derivative is

$$D_{DL\mu} = \partial/\partial X^{\mu} - \tfrac{1}{2} ig_D\boldsymbol{\sigma}\cdot\mathbf{W}'_{\mu} + \tfrac{1}{2} ig_D' B'_{\mu} + \tfrac{1}{2} ig_D'' B_{\mu} - \tfrac{1}{2} ig_G \mathbf{G}\cdot\mathbf{U}_{\mu} \quad (16.4)$$

with $\boldsymbol{\sigma}$ a vector composed of the Pauli matrices. The right-handed covariant derivatives have a simpler form. The normal matter right-handed covariant derivative is

$$D_{NR\mu} = \partial/\partial X^{\mu} + \tfrac{1}{2} ig' B_{\mu} - \tfrac{1}{2} ig_G \mathbf{G}\cdot\mathbf{U}_{\mu} \quad (16.5)$$

and the Dark matter right-handed covariant derivative is

$$D_{DR\mu} = \partial/\partial X^{\mu} + \tfrac{1}{2} ig_D' B'_{\mu} + \tfrac{1}{2} ig_D'' B_{\mu} - \tfrac{1}{2} ig_G \mathbf{G}\cdot\mathbf{U}_{\mu} \quad (16.6)$$

The normal and Dark electroweak fields above are functions of Two-Tier coordinates X. The Faddeev-Popov mechanism operative for these types of fields is described in appendix 19-A of Blaha (2011c) and in chapter 12.

16.1.2 Quark Sector

In the *quark* sector we define left and right quark quadruplets with

$$\Psi_{qL,Ra}(X_c) = \begin{bmatrix} \psi_{DqL,Ra}(X_c) \\ \psi_{NqL,Ra}(X_c) \end{bmatrix} \quad (16.7)$$

where $\psi_{NqL,Ra}(X_c)$ is a "normal" ElectroWeak-like quark doublet consisting of an SU(3) color up-quark and a color SU(3) down-quark, and where $\psi_{DqL,Ra}(X_c)$ is a Dark ElectroWeak-like quark doublet consisting of an SU(3) singlet Dark up-quark of unit Dark charge and an SU(3) singlet Dark down-quark of zero Dark charge in the ath generation.

The covariant derivative terms are contained in $D_q(X_c)$ which we express in matrix form as

$$D_{qL,R}(X_c) = \begin{bmatrix} \gamma^\mu D_{qDL,R\mu}(X_c) & 0 \\ 0 & \gamma^\mu D_{qNL,R\mu}(X_c) \end{bmatrix} \tag{16.8}$$

where the normal quark matter left-handed covariant derivative is

$$D_{qNL\mu} = \partial/\partial X_c{}^\mu - \tfrac{1}{2}ig\boldsymbol{\sigma}\cdot\mathbf{W}_\mu - ig'B_\mu/6 - \tfrac{1}{2}ig_G\mathbf{G}\cdot\mathbf{U}_\mu + ig_C\boldsymbol{\tau}\cdot A_{C\mu} \tag{16.9}$$

and where the Dark quark left-handed covariant derivative is

$$D_{qDL\mu} = \partial/\partial X_c{}^\mu - \tfrac{1}{2}ig_D\boldsymbol{\sigma}\cdot\mathbf{W}'_\mu + \tfrac{1}{2}ig_D'B'_\mu + \tfrac{1}{2}ig_D''B_\mu - \tfrac{1}{2}ig_G\mathbf{G}\cdot\mathbf{U}_\mu \tag{16.10}$$

since Dark quarks are SU(3) singlets with unit or zero Dark charge. The right-handed quark covariant derivatives have a simpler form. The normal quark right-handed covariant derivative is

$$D_{qNR\mu} = \partial/\partial X_c{}^\mu + \tfrac{1}{2}ig'B_\mu/3 - \tfrac{1}{2}ig_G\mathbf{G}\cdot\mathbf{U}_\mu + ig_C\boldsymbol{\tau}\cdot A_{C\mu} \tag{16.11}$$

and the Dark quark right-handed covariant derivative is

$$D_{qDR\mu} = \partial/\partial X_c{}^\mu + \tfrac{1}{2}ig_D'B'_\mu + \tfrac{1}{2}ig_D''B_\mu - \tfrac{1}{2}ig_G\mathbf{G}\cdot\mathbf{U}_\mu \tag{16.12}$$

The normal and Dark gauge boson fields are functions of the Two-Tier coordinates X_c. $= (X_{r\mu}(y_r), X_{i\mu}(y_i))$ of eqs. 14.11 and 14.12. The Faddeev-Popov mechanism is operative for gauge boson fields and is described in appendix 19-A of Blaha (2011c).[29] The *complexon* quark Extended Standard Model ElectroWeak Sector covariant derivatives in quadruplet matrix form are

$$D_{qL,R}(X_c) = \begin{bmatrix} \gamma^\mu D_{qDL,R\mu} & 0 \\ 0 & \gamma^\mu D_{qNL,R\mu} \end{bmatrix} \tag{16.13}$$

[29] Those who might be concerned about the propagator term $<W_i(X), W_j(X_c)>$ and similar propagators where one field is a function of X and the other field is a function of X_c should note that such terms are to very good approximation equal to $<W_i(X), W_j(X)>$ for energies much less than M_c. (These energies could be as large as the Planck energy.)

The remaining parts of the complexon Standard Model are described in chapter 23 of Blaha (2011) and summarized below. The addition of singlet Dark quark Higgs terms is also required.

The lagrangian density and action is

$$\mathcal{L}_{CSM} = \Psi_L^{\dagger}\gamma^0 i\gamma^{\mu}D_{L\mu}\Psi_L - \Psi_R^{\dagger}\gamma^0 i\gamma^{\mu}D_{R\mu}\Psi_{3R} + \Psi_{CL}^{\dagger}\gamma^0 i\gamma^{\mu}\mathcal{D}_{qL\mu}\Psi_{CL} + \Psi_{CR}^{\dagger}\gamma^0 i\gamma^{\mu}\mathcal{D}_{qR\mu}\Psi_{CR} - \\ - \mathcal{L}_{BareMasses} + \mathcal{L}_{Gauge} + \mathcal{L}_{Mass} + \mathcal{L}_{Ufields} \tag{16.14}$$

where there is an implicit sum over generations. $\mathcal{L}_{BareMasses}$ contains the fermion bare mass terms. Also,

$$\mathcal{L}_{Gauge} = \mathcal{L}_{GaugeEW} + \mathcal{L}_{GaugeC} + \mathcal{L}_{GaugeEWD} \tag{16.15}$$

with

$$\mathcal{L}_{GaugeEW} = -\tfrac{1}{4}\, F_W{}^{a\mu\nu}F_W{}^{a}{}_{\mu\nu} - \tfrac{1}{4}\, F_B{}^{\mu\nu}F_{B\mu\nu} + \mathcal{L}_{EW}{}^{ghost} \tag{16.16}$$

$$\mathcal{L}_{GaugeEWD} = -\tfrac{1}{4}\, F'_W{}^{a\mu\nu}F'_W{}^{a}{}_{\mu\nu} - \tfrac{1}{4}\, F_{B'}{}^{\mu\nu}F_{B'\mu\nu} + \mathcal{L}_{W'}{}^{ghost} \tag{16.17}$$

and

$$\mathcal{L}_{GaugeC} = \mathcal{L}_{CCG} + \mathcal{L}_C{}^{ghost} + \mathcal{L}_{CC}{}^{ghost} \tag{16.18}$$

$$\mathcal{L}_{Ufields} = -\tfrac{1}{4}\, F_U{}^{a\mu\nu}F_{U\mu\nu\nu} + \mathcal{L}_U{}^{ghost} + \mathcal{L}_U{}^{UHiggs} \tag{16.19}$$

where $\mathcal{L}_U{}^{UHiggs}$ is discussed in section 16.4. The ElectroWeak gauge bosons $W_\mu{}^a$, B_μ and B'_μ field tensors are:

$$F_W{}^a{}_{\mu\nu} = \partial W^a{}_\mu/\partial X^\nu - \partial W^a{}_\nu/\partial X^\mu + g_2 f^{abc} W^b{}_\mu W^c{}_\nu \tag{16.20}$$

$$F_{B\mu\nu} = \partial B_\mu/\partial X^\nu - \partial B_\nu/\partial X^\mu \tag{16.21}$$

and the Dark ElectroWeak gauge bosons $W'_\mu{}^a$ and B'_μ field tensors are:

$$F_{B'\mu\nu} = \partial B'_\mu/\partial X^\nu - \partial B'_\nu/\partial X^\mu \\ F'_W{}^a{}_{\mu\nu} = \partial W'^a{}_\mu/\partial X^\nu - \partial W'^a{}_\nu/\partial X^\mu + g_2 f^{abc} W'^b{}_\mu W'^c{}_\nu \tag{16.22}$$

The U fields' tensor is:

$$F_U{}^a{}_{\mu\nu} = \partial U^a{}_\mu/\partial X^\nu - \partial U^a{}_\nu/\partial X^\mu + g_G f_4{}^{abc} U^b{}_\mu U^c{}_\nu \tag{16.23}$$

where $f_4{}^{abc}$ are the U(4) algebra structure constants.

$\mathcal{L}_{EW}^{ghost}$ contains the Faddeev-Popov ghost terms for the ElectroWeak $W_\mu{}^a$ gauge bosons. The complexon color gluon lagrangian $\mathcal{L}_{CCG}$ is defined by

$$\mathcal{L}_{CCG} = -\tfrac{1}{4}\, F_{CC}{}^{a\mu\nu}(X) F_{CC}{}^{a}{}_{\mu\nu}(X) \tag{16.24}$$

where

$$F_{CC}{}^{a}{}_{\mu\nu} = \partial/\partial X_c{}^{\nu}\, A_C{}^{a}{}_{\mu} - \partial/\partial X_c{}^{\mu}\, A_C{}^{a}{}_{\nu} + g f_{su(3)}{}^{abc} A_C{}^{b}{}_{\mu} A_C{}^{c}{}_{\nu} \tag{16.25}$$

where $A_C{}^a{}_\nu$ is the color gluon gauge field, g is the color coupling constant, and the $f_{su(3)}{}^{abc}$ are the SU(3) structure constants.

In addition $\mathcal{L}_C^{ghost}$ is the color SU(3) Faddeev-Popov ghost terms defined in appendix 19-A of Blaha (2011c) for the complexon Lorentz gauge and $\mathcal{L}_{CC}^{ghost}$ is the complexon color SU(3) constraint ghost terms defined through the Faddeev-Popov mechanism. The mass sector $\mathcal{L}_{Mass}$ is presumably based on the Higgs Mechanism, which creates the fermion and ElectroWeak vector boson masses, and generation mixing.

The lagrangian is supplemented with the following condition on all complexon fields $\Phi_{...}$:[30]

$$\nabla_r \cdot \nabla_i \Phi ... = 0 \tag{16.26}$$

Non-complexon fields Ω... in the left-handed formulation under consideration satisfy the subsidiary condition:

$$[\nabla_r \cdot \nabla_i - (\nabla_r^2 \nabla_i^2)^{1/2}]\Omega ... = 0 \tag{16.27}$$

which guarantees a complexon's real 3-momentum is parallel to its imaginary 3-momentum.

16.2 Generation U(4) Gauge Symmetry Breaking and Long Range Forces

In chapter 15 we showed that there was good experimental evidence for a conserved Baryon Number B and we proceeded to develop a simple U(1) gauge theory that would imply Baryon Number conservation in a manner analogous to QED's implying electric charge conservation. In section 16.1 we used a new symmetry group local U(4) to generalize the one generation Extended Standard Model to a four generation Extended Standard Model based on four conserved particle numbers: B, L, B_D, and L_D.[31]

We now assume in our construction that the four generation Extended Standard Model has a local U(4) symmetry that is broken by mass terms generated by the Higgs Mechanism.

Further, we will assume that the Higgs breakdown yields two massless (long-range) fields, which we associate with Baryon Number B and Dark Baryon Number B_D. The remaining fields acquire masses and generate short-range forces.

---

[30] These conditions implement the orthogonality of the real and imaginary parts of complexon 3-momentum.

[31] Charge, although a conserved number, is a part of the ElectroWeak sector, account of which has already been taken.

We use the following U(4) diagonal matrices:

$$G_1 = \mathrm{diag}(1, 1, 1, 1)$$
$$G_2 = \mathrm{diag}(0, 1, 0, 0)$$
$$G_3 = \mathrm{diag}(0, 0, 1, 0)$$
$$G_4 = \mathrm{diag}(0, 0, 0, 1) \quad (16.28)$$

The U(4) algebra has 16 hermitean matrices that satisfy

$$G_i^{\dagger} = G_i \quad (16.29)$$

The particle numbers can be expressed in terms of the diagonal generators as

$$B = G_1 - G_2 - G_3 - G_4$$
$$B_D = G_2$$
$$L = G_3$$
$$L_D = G_4 \quad (16.30)$$

The covariant derivatives have the general form:

$$D_{\ldots\mu} = \partial/\partial X^{\mu} + \ldots - \tfrac{1}{2} i g_G \mathbf{G}\cdot\mathbf{U}_{\mu} \quad (16.31)$$

where the ellipsis's indicates the other details of the particular covariant derivative. We now wish to express the four gauge fields $U_i(X)$ for i = 1, 2, 3, 4 corresponding to the diagonal generators in terms of the fields of the four particle number gauge fields: B_{μ}, L_{μ}, $B_{D\mu}$, and $L_{D\mu}$.

$$U_{i\mu} = A_{ik} N_{k\mu} \quad (16.32)$$

where A_{ik} are the elements of a matrix of constants and

$$N_{\mu} = \begin{bmatrix} B_{\mu}(X) \\ L_{\mu}(X) \\ B_{D\mu}(X) \\ L_{D\mu}(X) \end{bmatrix} \quad (16.33)$$

is a column vector consisting of the gauge fields corresponding to each of the conserved particle numbers.

The matrix A must have non-zero determinant so that eq. 16.32 can be inverted to express the particle number fields in terms of the four $U_i(X)$ gauge fields:

$$N_{\mu} = A^{-1}U_{\mu} \quad (16.34)$$

resulting in

$$B_\mu(X) = U_{1\mu} \qquad (16.35)$$
$$L_\mu(X) = U_{1\mu} + U_{2\mu}$$
$$B_{D\mu}(X) = U_{1\mu} + U_{3\mu}$$
$$L_{D\mu}(X) = U_{1\mu} + U_{4\mu}$$

Then

$$D_{\ldots\mu} = \partial/\partial X^\mu + \ldots - \tfrac{1}{2} i g_G[\sum_{i=5}^{16} \mathbf{G}_i \mathbf{U}_{i\mu} + BB_\mu(X) + LL_\mu(X) + B_D B_{D\mu}(X) + L_D L_{D\mu}(X)] \qquad (16.36)$$

where the particle numbers, which are analogous to the charges Q and Q' in ElectroWeak theory, are B, L, B_D, and L_D. They are expressed in terms of U(4) generators by eqs. 16.30.

16.3 Higgs Mass Mechanism for Generation U(4) Gauge Fields

We now require that there are two massless fields: one coupled to Baryon number and one coupled to Dark Baryon number. The Dark sector is assumed to be analogous to the normal particle sector in this respect. Most of the fourteen remaining fields acquire masses and longitudinal components. These fields become short-range, ultra-weak generation forces. The masses they acquire through the Higgs Mechanism are presumably very large, as these gauge particles have not been found experimentally.[32]

We assume that a scalar Higgs field exists, which is a U(4) vector with four components corresponding to the fermion generations.[33] It is an SU(2)⊗U(1)⊗SU(3) ElectroWeak scalar. Its lagrangian density is

$$\mathcal{L}_U^{\ UHiggs} = (\partial\eta^\dagger/\partial X^\mu)(\partial\eta/\partial X^\mu) - \lambda(\eta^\dagger\eta - \rho^2)^2 + \mathcal{L}_U^{\ UHiggs}{}_{FermionMasses}$$

where $\mathcal{L}_U^{\ UHiggs}{}_{FermionMasses}$ are the fermion masses produced by the U Higgs Mechanism and where we choose a unitary gauge in which the vector η is

$$\eta = \begin{bmatrix} 0 \\ \rho_1 \\ 0 \\ \rho_2 \end{bmatrix} \qquad (16.37)$$

where ρ_1 and ρ_2 are Higgs fields with vacuum expectation values. Then the covariant derivative of η is

32 Section 16.4 discusses this topic in more detail.

33 We use the conventional Higgs particle formalism here because of its familiarity. Based on the preceding chapters the pseudoquantum formalism should be used where the vector $\eta = \varphi_1 + \varphi_2$ in an appropriately modified lagrangian. The non-zero vacuum expectation values of the φ_1 components are ρ_1 and ρ_2.

$$D_{\ldots\mu}\eta = \{\partial/\partial X^\mu + \ldots - \tfrac{1}{2}ig_G[\Sigma \mathbf{G}_i\mathbf{U}_{i\mu} + BB_\mu(X) + LL_\mu(X) + B_D B_{D\mu}(X) + L_D L_{D\mu}(X)]\} \begin{bmatrix} 0 \\ \rho_1 \\ 0 \\ \rho_2 \end{bmatrix} \tag{16.38}$$

The sum over i is from 5 through 16, and $[\mathbf{G}_i]_{jk}$ is the jk[th] element of $\mathbf{G}_i$. Then

$$D_{\ldots\mu}\eta = \begin{bmatrix} -\tfrac{1}{2}ig_G\{\rho_1\Sigma[\mathbf{G}_i]_{12}\mathbf{U}_{i\mu} + \rho_2\Sigma[\mathbf{G}_i]_{14}\mathbf{U}_{i\mu}\} \\ \partial\rho_1/\partial X^\mu - \tfrac{1}{2}ig_G\rho_1 L_\mu - \tfrac{1}{2}ig_G\{\rho_1\Sigma[\mathbf{G}_i]_{22}\mathbf{U}_{i\mu} + \rho_2\Sigma[\mathbf{G}_i]_{24}\mathbf{U}_{i\mu}\} \\ -\tfrac{1}{2}ig_G\{\rho_1\Sigma[\mathbf{G}_i]_{32}\mathbf{U}_{i\mu} + \rho_2\Sigma[\mathbf{G}_i]_{34}\mathbf{U}_{i\mu}\} \\ \partial\rho_2/\partial X^\mu - \tfrac{1}{2}ig_G\rho_2 L_{D\mu} - \tfrac{1}{2}ig_G\{\rho_1\Sigma[\mathbf{G}_i]_{42}\mathbf{U}_{i\mu} + \rho_2\Sigma[\mathbf{G}_i]_{44}\mathbf{U}_{i\mu}\} \end{bmatrix} \tag{16.39}$$

$$= \begin{bmatrix} -\tfrac{1}{2}ig_G\Sigma\{\rho_1[\mathbf{G}_i]_{12} + \rho_2[\mathbf{G}_i]_{14}\}\mathbf{U}_{i\mu} \\ \partial\rho_1/\partial X^\mu - \tfrac{1}{2}ig_G\rho_1 L_\mu - \tfrac{1}{2}ig_G\rho_2\Sigma[\mathbf{G}_i]_{24}\mathbf{U}_{i\mu} \\ -\tfrac{1}{2}ig_G\Sigma\{\rho_1[\mathbf{G}_i]_{32} + \rho_2[\mathbf{G}_i]_{34}\}\mathbf{U}_{i\mu} \\ \partial\rho_2/\partial X^\mu - \tfrac{1}{2}ig_G\rho_2 L_{D\mu} - \tfrac{1}{2}ig_G\rho_1\Sigma[\mathbf{G}_i]_{42}\mathbf{U}_{i\mu} \end{bmatrix} \tag{16.40}$$

since the generators $\mathbf{G}_i$ have zeroes along their diagonals for i = 5, … , 16.

From eq. 16.39 we find the corresponding Higgs field kinetic terms in the lagrangian are

$$(D_{\ldots\mu}\eta)^\dagger D_{\ldots}^{\ \mu}\eta = \partial\rho_1/\partial X^\mu\,\partial\rho_1/\partial X_\mu + \partial\rho_2/\partial X^\mu\,\partial\rho_2/\partial X_\mu + g_G^2\rho_1^2 L_\mu L^\mu/4 + g_G^2\rho_2^2 L_{D\mu} L_D^{\ \mu}/4 + \ldots \tag{16.41}$$

Note there are differing mass squared terms for the Lepton ($g_G^2\rho_1^2/4$) and Dark Lepton ($g_G^2\rho_2^2/4$) gauge fields making them short range fields with the likelihood of very large masses much beyond ElectroWeak gauge field masses, and with an ultra weak coupling constant g_G as suggested by the "experimental" coupling for the Baryonic force given in eq. 15.6.

The Baryonic and Dark Baryonic gauge fields are massless and thus long range although their coupling constant appears to be ultra weak – much below the gravitational coupling constant G.

We now turn to calculating the remaining terms in eq. 16.41 that determine the masses of the remaining 14 gauge fields. We begin by assigning matrix elements for the remaining hermitean U(4) generators:

$$[G_5]_{ik} = \delta_{i1}\delta_{k2} + \delta_{i2}\delta_{k1} \tag{16.42}$$
$$[G_6]_{ik} = -i\delta_{i1}\delta_{k2} + i\delta_{i2}\delta_{k1}$$

$$
\begin{aligned}
[G_7]_{ik} &= \delta_{i1}\delta_{k3} + \delta_{i3}\delta_{k1} \\
[G_8]_{ik} &= -i\delta_{i1}\delta_{k3} + i\delta_{i3}\delta_{k1} \\
[G_9]_{ik} &= \delta_{i1}\delta_{k4} + \delta_{i4}\delta_{k1} \\
[G_{10}]_{ik} &= -i\delta_{i1}\delta_{k4} + i\delta_{i4}\delta_{k1} \\
[G_{11}]_{ik} &= \delta_{i2}\delta_{k3} + \delta_{i3}\delta_{k2} \\
[G_{12}]_{ik} &= -i\delta_{i2}\delta_{k3} + i\delta_{i3}\delta_{k2} \\
[G_{13}]_{ik} &= \delta_{i2}\delta_{k4} + \delta_{i4}\delta_{k2} \\
[G_{14}]_{ik} &= -i\delta_{i2}\delta_{k4} + i\delta_{i4}\delta_{k2} \\
[G_{15}]_{ik} &= \delta_{i3}\delta_{k4} + \delta_{i4}\delta_{k3} \\
[G_{16}]_{ik} &= -i\delta_{i3}\delta_{k4} + i\delta_{i4}\delta_{k3}
\end{aligned}
$$

Then completing eq. 16.41 using eq. 16.40 we find

$$
\begin{aligned}
(D_{\ldots\mu}\,\eta)^\dagger D_{\ldots}^{\ \mu}\,\eta = {} & \partial\rho_1/\partial X^\mu \partial\rho_1/\partial X_\mu + \partial\rho_2/\partial X^\mu\, \partial\rho_2/\partial X_\mu + g_G^{\,2}\rho_1^{\,2}L_\mu\, L^\mu/4 + g_G^{\,2}\rho_2^{\,2}L_{D\mu}\, L_D^{\,\mu}/4 + \\
& + (g_G/2)^2\rho_1^{\,2}(U_5^{\,2} + U_6^{\,2}) + (g_G/2)^2\rho_2^{\,2}(U_9^{\,2} + U_{10}^{\,2}) + (g_G/2)^2\rho_1^{\,2}(U_{11}^{\,2} + U_{12}^{\,2}) + \\
& + (g_G/2)^2(\rho_1^{\,2} + \rho_2^{\,2})(U_{13}^{\,2} + U_{14}^{\,2}) + \ + (g_G/2)^2\rho_2^{\,2}(U_{15}^{\,2} + U_{16}^{\,2}) \qquad (16.43)
\end{aligned}
$$

up to total divergences, which generate surface terms which we discard, and assuming that all fields satisfy the gauge condition

$$
\partial U_i^{\,\mu}/\partial X^\mu = 0 \qquad (16.44)
$$

Note that there are no mass terms for $U_7(X)$ and $U_8(X)$ as well as $B_\mu(X)$ and $B_{D\mu}(X)$ due to our choice of unitary gauge eq. 16.37. Consequently there are four massless long range fields and 12 gauge fields that acquire masses of three different values: $(g_G/2)\rho_{10}$, $(g_G/2)\rho_{20}$, and $(g_G/2)(\rho_{10}^{\,2} + \rho_{20}^{\,2})^{1/2}$ where ρ_{10} and ρ_{20} ar the vacuum expectation values of ρ_1 and ρ_2 respectively.[34] The fields $U_7(X)$ and $U_8(X)$ are not "diagonal" and thus appear in the fermion sector as terms connecting fermions in different generations within the four species of normal fermions and within the four species of Dark fermions.[35] Therefore they do not change the values of any of the four types of particle numbers.

Based on the estimate of eq. 15.6 the ultra weak value of the coupling constant is

$$
g_G = (4\pi\alpha_B)^{1/2} \approx 1.218\ (Gm_H^{\,2})^{1/2} \qquad (16.45)
$$

The ultra-weak value of the coupling constant implies that the baryonic force with gauge field $B_\mu(X)$, which is now part of a quadruplet of fields, is a massless, long range field that corresponds to that of chapter 15 with the exception that chapter 15 looks ahead to later chapters where we discuss a 16-dimensional space that we call the *Megaverse* in which our universe resides where the baryonic force and force associated with the Dark Baryon force exist

---

[34] The Higgs fields ρ_i in our pseudoquantum formulation are $\rho_i = \varphi_{1i}(x) + \varphi_{2i}(x)$ with ρ_{i0} being the vacuum expectation of φ_{1i} as described earlier.

[35] Neutral lepton, charged lepton, up-type quark and down-type quark plus the four corresponding Dark species.

beyond our universe and act with other possible "island" universes. (The leptonic and Dark leptonic forces are short range and thus do not extend beyond our universe.)

The two non-diagonal long-range forces, being between different generations of a species and having an ultra-weak coupling, are not of great consequence because of the short lifetime of the higher generations of a species. Therefore, despite their long range, they have only the "shortest" time to exert an inter-generation force before a higher generation particle decays.

Since we expect the other massive fields to have very large masses (and thus very large Higgs field vacuum expectation values) and ultra-weak coupling they are not likely to be experimentally found for the foreseeable future.

16.4 Impact of this Generation U(4) Higgs Mechanism on Fermion Generation Masses

The fermion masses of the charged lepton, and the up-type quark, and down-type quark species' generations all show a rapid increase of mass with the generation. For example the u quark mass is a few MeV while the t quark (third generation) has a mass of about 170 GeV/c. The ratio of these masses is about 170,000. While one can account for this great difference by the judicious choices of Higgs' parameter values, when one considers the generational group and its associated numerical quantities: ultra-weak coupling, very large U particle masses – perhaps of the order of hundreds or thousands of GeV/c, and the corresponding very large Higgs particle vacuum expectation values in the U gauge field sector[36] then the differences in fermion masses within a species become more understandable and natural from a Leibniz Principle perspective. [The Layer group interactions add further mass terms, which also may be partly responsible for the large differences in mass between the generations of charged species. We discuss this possibility in more detail later.]

Thus the popular view that the ElectroWeak gauge field symmetry breaking occurs solely via ElectroWeak Higgs fields is not part of our Extended Standard Model unless the U(4) sector is removed. In our model there are two sets of contributions to fermion symmetry breaking: ElectroWeak Higgs particles symmetry breaking, and Generation group U(4) Higgs particles symmetry breaking [expended later to include a Layer group sector]. The Generation group causes each species to break into four generations.

[*The U(4) Generation group adds 12 more generators (and thus interaction terms) to the 4 ElectroWeak generators. The four generations requires a 4×4 matrix representation which we take to consist of a reducible* $\underline{3}\oplus\underline{1}$ *representation of SU(2)⊗U(1).*]

In the conventional Standard Model the breakup of species into generations is inserted "by hand." It is not a consequence of the existence of SU(2)⊗U(1) symmetry or symmetry breaking. In our approach the U(4) Generation group causes the appearance of generations. We

[36] They are not the Higgs particles of the SU(2)⊗U(1) ElectroWeak sector.

base the existence of the Generation group[37] on the four, conserved particle numbers. Leibniz' Principle and Ockham's Razor then lead to the above construction/derivation.

16.5 Generation Group Higgs Mechanism for Fermion Masses

We now consider the Generation group Higgs Mechanism for the eight species of fermions (four species of "normal" matter[38] and four species of Dark Matter). We shall consider the mass terms for the four normal species, which is the same as that of the four Dark species except for the values in the various species mass matrices. Therefore we define the initial 4-vector for the generations of the normal species by

$$\Psi_s = \begin{bmatrix} \psi_{11} \\ \psi_{12} \\ \psi_{13} \\ \psi_{14} \\ \dots \\ \psi_{41} \\ \psi_{42} \\ \psi_{43} \\ \psi_{44} \end{bmatrix} \tag{16.46}$$

where ψ_{ki} is the generation index for the ith generation of the kth species. ψ_{k1} is the wave function for the 1st generation, ψ_{k4} is the 4th generation member of the kth species, and we omit other indices in the interests of clarity. The normal fermion species order here is: charged lepton (k = 1), up-type quark, neutral lepton, and down-type quark (k = 4). Other indices of these wave functions are suppressed in the interests of clarity. A 4th generation fermion of any species is yet to be found experimentally. The lagrangian density mass terms for the four normal fermion species are

$$\mathcal{L}_U{}^{UHiggs}{}_{FermionMasses} = \Sigma_{k,\alpha,\beta}\ \bar{\psi}_{kL\alpha}\ \eta_k m_{k\alpha\beta}\psi_{kR\beta}\ + \text{c.c.} \tag{16.47}$$

where $m_{k\alpha\beta}$ is complex constant matrix, where k labels species, and where $\alpha, \beta = 1, \dots, 4$. The total of fermion lagrangian mass terms are

$$\mathcal{L}^{Higgs}{}_{FermionMasses} = \mathcal{L}_U{}^{UHiggs}{}_{FermionMasses} + \mathcal{L}_{EW}{}^{Higgs}{}_{FermionMasses} \tag{16.48}$$

---

[37] In earlier books we suggested the fermion generations might be the result of a wormhole to another 4-dimensional universe. The new approach is simpler and more consistent with known facts – thus more consistent with the Leibniz Minimax Principle.

[38] Not taking account of the three color quark species of normal matter yet.

where $\mathcal{L}_{EW}^{Higgs}$ is the contribution of ElectroWeak Higgs Mechanism to the fermion masses (discussed in the following chapter). Using the vacuum expectation value of η in eq. 16.37 we find

$$\mathcal{L}_{U}^{UHiggs}{}_{FermionMasses} = \Sigma_{\alpha,\beta}\ \{\bar{\psi}\psi_{2L\alpha}\ \rho_1 m_{2\alpha\beta}\psi_{2R\beta} + \bar{\psi}_{4L\alpha}\ \rho_2 m_{4\alpha\beta}\psi_{4R\beta}\} + \text{c.c.} \tag{16.49}$$

giving mass terms for the up-type and down-type quark species but not for lepton species. There is an implicit color summation over the color quarks in each generation and quark species. *Qualitatively eq. 16.49 could be viewed as corresponding to the experimentally known largeness of quark masses relative to lepton masses in each generation of normal matter.*

The mass matrices $m_2 = [m_{2\alpha\beta}]$ and $m_4 = [m_{4\alpha\beta}]$ are both complex, constant mass matrices. They can be brought to diagonal form with non-negative values by U(4) matrices A_k and B_k:

$$A_2 m_2 B_2^{-1} = D_2 \qquad A_4 m_4 B_4^{-1} = D_4 \tag{16.50}$$

or

$$m_2 = A_2^{-1} D_2\, B_2 \qquad m_4 = A_4^{-1} D_4\, B_4 \tag{16.51}$$

We now note, that although, both D_2 and D_4 have non-negative real values, down-type quarks are all tachyonic and up-type quarks are all non-tachyonic due to their lagrangian kinetic terms as seen in chapter 5.

We further note that $m_2^\dagger m_2$ and $m_4^\dagger m_4$ are hermitean, and A_k and B_k are members of U(4) as is D_k for k = 2,4, with the result that m_2 and m_4 are also both members of the U(4) group. Thus

$$m_2^{-1} = m_2^\dagger \qquad m_4^{-1} = m_4^\dagger \tag{16.52}$$

We can express the mass matrices in terms of U(4) generators

$$m_2 = \Sigma G_i m_{2i} \qquad m_4 = \Sigma G_i m_{4i} \tag{16.53}$$

$$m_2^{-1} = m_2^\dagger = \Sigma G_i m_{2i}^* \qquad m_4^{-1} = m_4^\dagger = \Sigma G_i m_{4i}^* \tag{16.54}$$

since the matrices G_i are all hermitean, where $\{m_{2i}\}$ and $\{m_{4i}\}$ are each a set of sixteen complex constants.

While we do not as yet know the 4th generation fermions or their masses, the third generation quarks have masses that are far greater than the 1st and 2nd generation quarks or their sum suggesting that the trace of m_2 and m_4.is dominated by the 4th generation mass of the two quark species with a similar situation holding, perhaps, for the two Dark quark species. Therefore if we take the trace of m_2 and m_4 then it seems probable based on the trend of the generations that the 4th generation mass dominates the trace:

$$\begin{aligned} D_{24} &\approx \text{tr } D_2 \\ D_{44} &\approx \text{tr } D_4 \end{aligned} \tag{16.55}$$

We can use these A_k and B_k U(4) transformations to define the eight "physical" (up to further ElectroWeak Higgs Mehanism effects) up-type and down-type quark generations fields:

$$\begin{aligned} \bar{\psi}_{2L\alpha}\,\rho_1 m_{2\alpha\beta}\psi_{2R\beta} + \bar{\psi}_{4L\alpha}\,\rho_2 m_{4\alpha\beta}\psi_{4R\beta} &= (\bar{\psi}_{2L}A_2^{-1})_\alpha \rho_1 D_{2\alpha\beta}(B_2\psi_{2R})_\beta + (\bar{\psi}_{4L}A_4^{-1})_\alpha \rho_2 D_{4\alpha\beta}(B_4\psi_{4R})_\beta \\ &= \bar{\psi}_{2Lphys\alpha}\,\rho_1 D_{2\alpha\beta}\psi_{2Rphys\beta} + \bar{\psi}_{4Lphs\alpha}\,\rho_2 D_{4\alpha\beta}\psi_{4Rphys\beta} \end{aligned} \tag{16.56}$$

Species: up-type quarks down-type quarks

The preceding discussion with changes in the values of constants and constant matrices holds for Dark Matter also where the Dark quarks acquire mass terms but the Dark leptons do not. The Dark Matter species mass terms, with the subscript D signifying Dark Matter, are

$$= \bar{\psi}_{D2Lphys\alpha}\,\rho_{D1} D_{D2\alpha\beta}\psi_{D2Rphys\beta} + \bar{\psi}_{D4Lphs\alpha}\,\rho_{D2} D_{D4\alpha\beta}\psi_{D4Rphys\beta} \tag{16.57}$$

Dark Species: up-type quarks down-type quarks

END OF EXTRACT

4.4 Layer Group Interaction

In this section we describe Layer group interactions.[39] The Theory of Everything symmetry group that we have developed in earlier books was:

$$SU(3)\otimes SU(2)\otimes U(1)\otimes SU(2)\otimes U(1)\otimes U(4)\otimes U(4)) \tag{4.1}$$

[39] Much of this section is extracted from Blaha (2016b).

Each factor in this product is a subgroup of U(8) but the product of the subgroups is *not* a subgroup of U(8) because the elements of each factor does not commute with the elements of other factor groups. The total number of independent generators of the product

$$SU(3)\otimes SU(2)\otimes U(1)\otimes SU(2)\otimes U(1)\otimes U(4)\otimes U(4)$$

is 48. The total number of generators of U(8) is 64. The difference, 16 generators, constitutes the generators of another U(4) group. We therefore propose that this U(4) group, which we call the Layer group, be a part of the Theory of Everything group yielding:

$$SU(3)\otimes SU(2)\otimes U(1)\otimes SU(2)\otimes U(1)\otimes U(4)\otimes U(4)\otimes U(4) \tag{4.2}$$

Its role will be to provide an ultra-weak interaction between the known normal fermions and the Dark fermions. In addition the Layer group also naturally leads to four layers of fermions. Our normal matter and its associated Dark Matter we call Layer 1. The other three layers, with presumably very high masses, are Dark Matter also and remain to be discovered.

We now modify the leptonic left-handed and right-handed covariant derivatives in the normal and Dark ElectroWeak sectors, which were:[40]

$$D_{NL\mu} = \partial/\partial X^\mu - \tfrac{1}{2}ig\boldsymbol{\sigma}\cdot\mathbf{W}_\mu + \tfrac{1}{2}ig'B_\mu - \tfrac{1}{2}ig_G\mathbf{G}\cdot\mathbf{U}_\mu \tag{4.3}$$
$$D_{DL\mu} = \partial/\partial X^\mu - \tfrac{1}{2}ig_D\boldsymbol{\sigma}\cdot\mathbf{W}'_\mu + \tfrac{1}{2}ig_D'B'_\mu + \tfrac{1}{2}ig_D''B_\mu - \tfrac{1}{2}ig_G\mathbf{G}\cdot\mathbf{U}_\mu \tag{4.4}$$
$$D_{NR\mu} = \partial/\partial X^\mu + ig'B_\mu - \tfrac{1}{2}ig_G\mathbf{G}\cdot\mathbf{U}_\mu \tag{4.5}$$
$$D_{DR\mu} = \partial/\partial X^\mu + \tfrac{1}{2}ig_D'B'_\mu + \tfrac{1}{2}ig_D''B_\mu - \tfrac{1}{2}ig_G\mathbf{G}\cdot\mathbf{U}_\mu \tag{4.6}$$

where the B_μ term in eqs. 4.4 and 4.6 above provide a transition between normal and dark matter. We also now modify eqs. 4,8 and 4.10 in the quark ElectroWeak covariant derivatives, which were:

$$D_{qNL\mu} = \partial/\partial X_c^\mu - \tfrac{1}{2}ig\boldsymbol{\sigma}\cdot\mathbf{W}_\mu - ig'B_\mu/6 + ig_C\boldsymbol{\tau}\cdot A_{C\mu} - \tfrac{1}{2}ig_G\mathbf{G}\cdot\mathbf{U}_\mu \tag{4.7}$$
$$D_{qDL\mu} = \partial/\partial X_c^\mu - \tfrac{1}{2}ig_D\boldsymbol{\sigma}\cdot\mathbf{W}'_\mu + \tfrac{1}{2}ig_D'B'_\mu + \tfrac{1}{2}ig_D''B_\mu - \tfrac{1}{2}ig_G\mathbf{G}\cdot\mathbf{U}_\mu \tag{4.8}$$
$$D_{qNR\mu} = \partial/\partial X_c^\mu + ig'B_\mu/3 + ig_C\boldsymbol{\tau}\cdot A_{C\mu} - \tfrac{1}{2}ig_G\mathbf{G}\cdot\mathbf{U}_\mu \tag{4.9}$$
$$D_{qDR\mu} = \partial/\partial X_c^\mu + \tfrac{1}{2}ig_D'B'_\mu + \tfrac{1}{2}ig_D''B_\mu - \tfrac{1}{2}ig_G\mathbf{G}\cdot\mathbf{U}_\mu \tag{4.10}$$

where $A_{C\mu}$ is the color gauge field.

The new covariant derivatives that contain the U(4) Layer group interaction are:

1) Leptonic covariant derivatives in the normal and Dark ElectroWeak sectors:

$$D_{NL\mu} = \partial/\partial X^\mu - \tfrac{1}{2}ig\boldsymbol{\sigma}\cdot\mathbf{W}_\mu + \tfrac{1}{2}ig'B_\mu - \tfrac{1}{2}i\, g_v\mathbf{G}_L\cdot\mathbf{V}_\mu - \tfrac{1}{2}ig_G\mathbf{G}\cdot\mathbf{U}_\mu \tag{4.3'}$$
$$D_{DL\mu} = \partial/\partial X^\mu - \tfrac{1}{2}ig_D\boldsymbol{\sigma}\cdot\mathbf{W}'_\mu + \tfrac{1}{2}ig_D'B'_\mu - \tfrac{1}{2}i\, g_v\mathbf{G}_L\cdot\mathbf{V}_\mu - \tfrac{1}{2}ig_G\mathbf{G}\cdot\mathbf{U}_\mu \tag{4.4'}$$
$$D_{NR\mu} = \partial/\partial X^\mu + ig'B_\mu - \tfrac{1}{2}i\, g_v\mathbf{G}_L\cdot\mathbf{V}_\mu - \tfrac{1}{2}ig_G\mathbf{G}\cdot\mathbf{U}_\mu \tag{4.5'}$$

[40] Equations from Blaha (2015a).

$$D_{DR\mu} = \partial/\partial X^{\mu} + ½ig_D B'_{\mu} - ½i\, g_v \mathbf{G}_L\cdot\mathbf{V}_{\mu} - ½ig_G \mathbf{G}\cdot\mathbf{U}_{\mu} \quad (4.6')$$

2) Quark ElectroWeak covariant derivatives in the normal and Dark ElectroWeak sectors:

$$D_{qNL\mu} = \partial/\partial X_c^{\mu} - ½ig\boldsymbol{\sigma}\cdot\mathbf{W}_{\mu} - ig' B_{\mu}/6 + ig_C \boldsymbol{\tau}\cdot A_{C\mu} - ½i\, g_v \mathbf{G}_L\cdot\mathbf{V}_{\mu} - ½ig_G \mathbf{G}\cdot\mathbf{U}_{\mu} \quad (4.7')$$
$$D_{qDL\mu} = \partial/\partial X_c^{\mu} - ½ig_D\boldsymbol{\sigma}\cdot\mathbf{W}'_{\mu} + ½ig_D B'_{\mu} - ½i\, g_v \mathbf{G}_L\cdot\mathbf{V}_{\mu} - ½ig_G \mathbf{G}\cdot\mathbf{U}_{\mu} \quad (4.8')$$
$$D_{qNR\mu} = \partial/\partial X_c^{\mu} + ig' B_{\mu}/3 + ig_C \boldsymbol{\tau}\cdot A_{C\mu} - ½i\, g_v \mathbf{G}_L\cdot\mathbf{V}_{\mu} - ½ig_G \mathbf{G}\cdot\mathbf{U}_{\mu} \quad (4.9')$$
$$D_{qDR\mu} = \partial/\partial X_c^{\mu} + ½ig_D B'_{\mu} - ½i\, g_v \mathbf{G}_L\cdot\mathbf{V}_{\mu} - ½ig_G \mathbf{G}\cdot\mathbf{U}_{\mu} \quad (4.10')$$

We add a new Layer group index to each fermion field[41] with index number values ranging from 1 through 4. The new Layer group interaction term $g_v G_L{}^a V^a{}_{\mu}$ uses 16 U(4) gauge fields $V^a{}_{\mu}$ with a = 1, 2, …, 16, and 16 U(4) generators, denoted $G_L{}^a$, *that couple to the new Layer group indexes.*[42] The $V^a{}_{\mu}$ gauge fields have a standard U(4) kinetic energy lagrangian term.

The Layer interaction constant g_v is an ultra-weak coupling constant assumed to be much smaller than the ElectroWeak and Generation group coupling constants.

In addition all gauge vector bosons acquire a Layer group index appropriate to the layer and all Higgs bosons of each layer have a layer index appropriate to the layer. Thus each layer is effectively self-contained,[43] with different fermion and boson particle masses, with only the ultra-weak Layer interaction and gravitation coupling layers.

4.5 Layer Group Higgs Mechanism Contributions to Layer Gauge Field Masses

In this section we will determine the Layer group Higgs contributions to gauge field masses. (The fermion mass contributions from the various Higgs interactions are shown in eq. 5.56' in chapter 3. We will see that all[44] layers have Layer group Higgs contributions to their fermion masses.)

We begin by assuming that a scalar Higgs field η exists, which is a U(4) Layer group 4-vector with four components corresponding to the four conserved generation number operators L_1, L_2, L_3, and L_4 of section 4.1. η is an SU(2)⊗U(1)⊗SU(3) ElectroWeak and Strong Interaction scalar. Its lagrangian density terms are[45]

$$\mathcal{L}_V{}^{Higgs} = (\partial\eta^{\dagger}/\partial X^{\mu})(\partial\eta/\partial X^{\mu}) - \lambda(\eta^{\dagger}\eta - \rho^2)^2 + \mathcal{L}_V{}^{Higgs}{}_{FermionMasses} \quad (4.11)$$

---

[41] Gauge bosons and Higgs bosons also acquire Layer group indexes. Thus each layer is entirely self-contained with the only interactions between layers being Layer group interactions and gravitation.

[42] The Generation group U(4) matrices G^a couple to Generation group indices.

[43] Modulo mass matrix mixing between the layers that will modify the fermion spectrum and gauge fields. These modifications are assumed to be very small due to the ultra-weak nature of the Layer group interaction.

[44] All Layers have Layer group Higgs contributions is required to avoid massless Layer group gauge fields.

[45] Again we use the standard formulation of the Higgs Mechanism because of its familiarity.

where $\mathcal{L}_V{}^{Higgs}{}_{FermionMasses}$ are the fermion masses produced by the Layer Higgs Mechanism and where we set the η Layer 4-vector with Higgs field components to

$$\eta = \begin{bmatrix} \rho_1 \\ \rho_2 \\ \rho_3 \\ \rho_4 \end{bmatrix} \qquad \begin{matrix} \underline{\text{Corresponding Conserved Number}} \\ L_1 \\ L_2 \\ L_3 \\ L_4 \end{matrix} \tag{4.12}$$

where ρ_1, ρ_2, ρ_3 and ρ_4 are real fields.[46] Then the covariant derivative of η is

$$D_{\ldots\mu}\eta = \{\partial/\partial X^\mu + \ldots - \tfrac{1}{2}ig_V[\Sigma\mathbf{G}_{Li}\mathbf{V}_{i\mu} + \mathbf{G}_{L1}\mathbf{V}_{1\mu} + \mathbf{G}_{L2}\mathbf{V}_{2\mu} + \mathbf{G}_{L3}\mathbf{V}_{3\mu} + \mathbf{G}_{L4}\mathbf{V}_{4\mu}]\} \begin{bmatrix} \rho_1 \\ \rho_2 \\ \rho_3 \\ \rho_4 \end{bmatrix} \tag{4.13}$$

The sum over i is from 5 through 16 (non-diagonal matrices), and $[\mathbf{G}_{Li}]_{jk}$ is the jk[th] element of $\mathbf{G}_{Li}$. Then

$$D_{\ldots\mu}\eta = \begin{bmatrix} \partial\rho_1/\partial X^\mu - \tfrac{1}{2}ig_V\{\rho_1\mathbf{G}_{L1}\mathbf{V}_{1\mu} + \rho_2\Sigma[\mathbf{G}_{Li}]_{11}\mathbf{V}_{i\mu} + \rho_2\Sigma[\mathbf{G}_{Li}]_{12}\mathbf{V}_{i\mu} + \rho_3\Sigma[\mathbf{G}_{Li}]_{13}\mathbf{V}_{i\mu} + \rho_4\Sigma[\mathbf{G}_{Li}]_{14}\mathbf{V}_{i\mu}\} \\ \partial\rho_2/\partial X^\mu - \tfrac{1}{2}ig_V\{\rho_2\mathbf{G}_{L2}\mathbf{V}_{2\mu} + \rho_1\Sigma[\mathbf{G}_{Li}]_{21}\mathbf{V}_{i\mu} + \rho_2\Sigma[\mathbf{G}_{Li}]_{22}\mathbf{V}_{i\mu} + \rho_3\Sigma[\mathbf{G}_{Li}]_{23}\mathbf{V}_{i\mu} + \rho_4\Sigma[\mathbf{G}_{Li}]_{24}\mathbf{V}_{i\mu}\} \\ \partial\rho_3/\partial X^\mu - \tfrac{1}{2}ig_V\{\rho_3\mathbf{G}_{L3}\mathbf{V}_{3\mu} + \rho_1\Sigma[\mathbf{G}_{Li}]_{31}\mathbf{V}_{i\mu} + \rho_2\Sigma[\mathbf{G}_{Li}]_{32}\mathbf{V}_{i\mu} + \rho_3\Sigma[\mathbf{G}_{Li}]_{33}\mathbf{V}_{i\mu} + \rho_4\Sigma[\mathbf{G}_{Li}]_{34}\mathbf{V}_{i\mu}\} \\ \partial\rho_4/\partial X^\mu - \tfrac{1}{2}ig_V\{\rho_4\mathbf{G}_{L4}\mathbf{V}_{4\mu} + \rho_1\Sigma[\mathbf{G}_{Li}]_{41}\mathbf{V}_{i\mu} + \rho_2\Sigma[\mathbf{G}_{Li}]_{42}\mathbf{V}_{i\mu} + \rho_3\Sigma[\mathbf{G}_{Li}]_{43}\mathbf{V}_{i\mu} + \rho_4\Sigma[\mathbf{G}_{Li}]_{44}\mathbf{V}_{i\mu}\} \end{bmatrix} \tag{4.14}$$

$$= \begin{bmatrix} \partial\rho_1/\partial X^\mu - \tfrac{1}{2}ig_V\{\rho_1\mathbf{G}_{L1}\mathbf{V}_{1\mu} + \rho_2\Sigma[\mathbf{G}_{Li}]_{12}\mathbf{V}_{i\mu} + \rho_3\Sigma[\mathbf{G}_{Li}]_{13}\mathbf{V}_{i\mu} + \rho_4\Sigma[\mathbf{G}_{Li}]_{14}\mathbf{V}_{i\mu}\} \\ \partial\rho_2/\partial X^\mu - \tfrac{1}{2}ig_V\{\rho_2\mathbf{G}_{L2}\mathbf{V}_{2\mu} + \rho_1\Sigma[\mathbf{G}_{Li}]_{21}\mathbf{V}_{i\mu} + \rho_3\Sigma[\mathbf{G}_{Li}]_{23}\mathbf{V}_{i\mu} + \rho_4\Sigma[\mathbf{G}_{Li}]_{24}\mathbf{V}_{i\mu}\} \\ \partial\rho_3/\partial X^\mu - \tfrac{1}{2}ig_V\{\rho_3\mathbf{G}_{L3}\mathbf{V}_{3\mu} + \rho_1\Sigma[\mathbf{G}_{Li}]_{31}\mathbf{V}_{i\mu} + \rho_2\Sigma[\mathbf{G}_{Li}]_{32}\mathbf{V}_{i\mu} + \rho_4\Sigma[\mathbf{G}_{Li}]_{34}\mathbf{V}_{i\mu}\} \\ \partial\rho_4/\partial X^\mu - \tfrac{1}{2}ig_V\{\rho_4\mathbf{G}_{L4}\mathbf{V}_{4\mu} + \rho_1\Sigma[\mathbf{G}_{Li}]_{41}\mathbf{V}_{i\mu} + \rho_2\Sigma[\mathbf{G}_{Li}]_{42}\mathbf{V}_{i\mu} + \rho_3\Sigma[\mathbf{G}_{Li}]_{43}\mathbf{V}_{i\mu}\} \end{bmatrix} \tag{4.15}$$

since the generators $\mathbf{G}_i$ have zeroes along their diagonals for i = 5, … , 16.

From eq. 4.15 we find the corresponding Higgs field kinetic terms in the lagrangian are

$$(D_{\ldots\mu}\eta)^\dagger D_{\ldots}{}^{\mu}\eta = \partial\rho_1/\partial X^\mu\,\partial\rho_1/\partial X_\mu + \partial\rho_2/\partial X^\mu\,\partial\rho_2/\partial X_\mu + \partial\rho_3/\partial X^\mu\,\partial\rho_3/\partial X_\mu + \partial\rho_4/\partial X^\mu\,\partial\rho_4/\partial X_\mu + $$
$$+ g_V^2\rho_1^2\mathbf{V}_{1\mu}\mathbf{V}_1{}^\mu/4 + g_V^2\rho_2^2\mathbf{V}_{2\mu}\mathbf{V}_2{}^\mu/4 + g_V^2\rho_3^2\,\mathbf{V}_{3\mu}\mathbf{V}_3{}^\mu/4 + g_V^2\rho_4^2\,\mathbf{V}_{4\mu}\mathbf{V}_4{}^\mu/4 + \ldots \tag{4.16}$$

[46] Each field ρ_i can be expressed as a pseudoquantum field: $\rho_i = \varphi_{1i} + \varphi_{2i}$ where φ_{1i} has the vacuum expectation value $\rho_{i0.\ \text{for } i = 1, \ldots, 4.}$ Thus our pseudoquantum field theory version is implemented easily.

We now turn to calculating the remaining terms in eq. 4.16 that determine the masses of the remaining 14 gauge fields. We begin by assigning matrix elements for the remaining hermitean U(4) generators:

$$\begin{aligned}
[G_{L5}]_{ik} &= \delta_{i1}\delta_{k2} + \delta_{i2}\delta_{k1} \\
[G_{L6}]_{ik} &= -i\delta_{i1}\delta_{k2} + i\delta_{i2}\delta_{k1} \\
[G_{L7}]_{ik} &= \delta_{i1}\delta_{k3} + \delta_{i3}\delta_{k1} \\
[G_{L8}]_{ik} &= -i\delta_{i1}\delta_{k3} + i\delta_{i3}\delta_{k1} \\
[G_{L9}]_{ik} &= \delta_{i1}\delta_{k4} + \delta_{i4}\delta_{k1} \\
[G_{L10}]_{ik} &= -i\delta_{i1}\delta_{k4} + i\delta_{i4}\delta_{k1} \\
[G_{L11}]_{ik} &= \delta_{i2}\delta_{k3} + \delta_{i3}\delta_{k2} \\
[G_{L12}]_{ik} &= -i\delta_{i2}\delta_{k3} + i\delta_{i3}\delta_{k2} \\
[G_{L13}]_{ik} &= \delta_{i2}\delta_{k4} + \delta_{i4}\delta_{k2} \\
[G_{L14}]_{ik} &= -i\delta_{i2}\delta_{k4} + i\delta_{i4}\delta_{k2} \\
[G_{L15}]_{ik} &= \delta_{i3}\delta_{k4} + \delta_{i4}\delta_{k3} \\
[G_{L16}]_{ik} &= -i\delta_{i3}\delta_{k4} + i\delta_{i4}\delta_{k3}
\end{aligned} \tag{4.17}$$

Then completing eq. 4.16 using eq. 4.15 we find

$$\begin{aligned}
(D_{\dots\mu}\eta)^\dagger D_{\dots}^{\ \mu}\eta = {} & \partial\rho_1/\partial X^\mu\, \partial\rho_1/\partial X_\mu + \partial\rho_2/\partial X^\mu\, \partial\rho_2/\partial X_\mu + \partial\rho_3/\partial X^\mu\, \partial\rho_3/\partial X_\mu + \partial\rho_4/\partial X^\mu\, \partial\rho_4/\partial X_\mu + \\
& + g_V^2\rho_1^2 \mathbf{V}_{1\mu}\mathbf{V}_1^{\ \mu}/4 + g_V^2\rho_2^2\, V_2^2/4 + g_V^2\rho_3^2\, V_3^2/4 + g_V^2\rho_4^2\, V_4^2/4 + \\
& + (g_V/2)^2(\rho_1^2 + \rho_2^2)(V_5^2 + V_6^2) + (g_V/2)^2(\rho_1^2 + \rho_3^2)(V_7^2 + V_8^2) + \\
& + (g_V/2)^2(\rho_1^2 + \rho_4^2)(V_9^2 + V_{10}^2) + (g_V/2)^2(\rho_2^2 + \rho_3^2)(V_{11}^2 + V_{12}^2) + \\
& + (g_V/2)^2(\rho_2^2 + \rho_4^2)(V_{13}^2 + V_{14}^2) + (g_V/2)^2(\rho_3^2 + \rho_4^2)(V_{15}^2 + V_{16}^2)
\end{aligned}$$

4.18)

up to total divergences, which generate surface terms which we discard, and also assuming that all fields satisfy the gauge condition

$$\partial V_i^{\ \mu}/\partial X^\mu = 0 \tag{4.19}$$

Eq. 4-18 shows all Layer group gauge fields have masses. The combination of an ultra-weak coupling constant and very large gauge field masses results in extremely weak interactions between the fields in each layer, which leads to almost independent layers of normal and Dark fermions. Thus the Darkness! They result in very rare decays between layers, and very weak interactions between fermions in different layers. The higher layers with presumably much more massive fermions are thus well "insulated" from our layer. Thus they are Dark to us as well.

We estimate Layer group gauge field masses to be very large – of the order of many TeV or they would have been detected at CERN by now. Their detection must await the construction of much more powerful accelerators. *The "non-diagonal" Layer gauge fields are the means by which we may hope to eventually find fermions of the higher layers.*

4.6 Layer Group Higgs Mechanism Contributions to Fermion Masses

The fermion masses of the charged lepton, and the up-type quark, and down-type quark species' generations all show a rapid increase of mass with the generation. For example the u quark mass is a few MeV while the t quark (third generation) has a mass of about 170 GeV/c. The ratio of these masses is about 170,000. While one can account for this great difference by the judicious choices of Higgs' parameter values, when one considers the Layer group and its associated numerical quantities: ultra-weak coupling, its very large Layer gauge field masses – perhaps of the order of hundreds or thousands of GeV/c, then a large difference in particle masses between layers is understandable and natural.

The form of the layers of fermion mass terms is[47]

$$\mathcal{L}^{\text{Higgs}}_{\text{FermionMasses}} = \Sigma_{k,a,\alpha,\beta}\bar{\psi}_{kaL\alpha\delta}\,\eta_k m_{EW_{ka\alpha\beta}}\psi_{kaR\beta} + \Sigma_{k,a,\alpha,\beta}\bar{\psi}_{D_{kaL\alpha}}\eta_{Dk} m_{DEW_{ka\alpha\beta}}\psi_{D_{kaR\beta}} + \qquad \text{ElectroWeak}$$

$$+ \Sigma_{k,a,\alpha,\beta}\bar{\psi}_{UkaL\alpha}\,\eta_{Uka} m_{Uka\alpha\beta}\psi_{UkaR\beta} + \qquad \text{Generation}$$

$$+ \Sigma_{k,a,\alpha,\beta}\bar{\psi}_{DUkaL\alpha}\eta_{DUka} m_{DUka\alpha\beta}\psi_{DUkaR\beta} + \qquad \text{Group U}$$

$$+ \Sigma_{k,g,\delta,\gamma}\bar{\psi}_{LkgL\delta}\,\eta_{Lg} m_{Lg\delta\gamma}\psi_{LkgR\gamma} + \qquad \text{Layer}$$

$$+ \Sigma_{k,g,\delta,\gamma}\bar{\psi}_{DLkgL\delta}\,\eta_{DLg} m_{DLg\delta\gamma}\psi_{DLkgR\gamma} + \qquad \text{Group L}$$

$$+ \Sigma_{k,a}\bar{\psi}_{GkaL}\eta_{Ga} m_{Gka}\psi_{GkaR} + \Sigma_{k,a}\bar{\psi}_{DGkaL}\eta_{DGa} m_{DGka}\psi_{DGkaR} + \qquad \text{Gravitational}$$

$$+ \text{c.c.} \tag{5.56'}$$

where the subscripts EW, D, U, L and G label ElectroWeak origin, D Dark type, U Generation group origin, L Layer group origin, and G Gravitational origin respectively. The fields labeled η (with subscripts) are Higgs fields that have non-zero vacuum expectation values.[48] The indices k label species – normal and Dark separately, g labels the (four) generations, and a labels the layers. The indices δ and γ label *layer* rows and columns (with implicit sums over generations in the Layer group terms.) The Layer group mass contribution is the same for each fermion in each generation for each species in each layer. The matrices labeled m (with subscripts) are the complex constant mass matrices of species. The indices α, β = 1, ... , 4 label *generation* rows and columns.

Eq. 5.56' contains the mass terms for the four layers of fermions in our Theory of Everything. *For each species and generation the Layer group terms mix the Layer mass contributions.*

---

[47] Layer group contributions have been added to the original eq. 5.56 in Blaha (2015b) in accord with Blaha (2016a).
[48] The Higgs fields η... in our pseudoquantum formulation are $\eta... = \varphi_{1...}(x) + \varphi_{2...}(x)$ as described earlier.

$$\mathcal{L}_V{}^{Higgs}{}_{FermionMasses} = \Sigma_{k,g,\delta,\gamma}\bar{\psi}_{LkgL\delta}\,\eta_{Lg}\,m_{Lg\delta\gamma}\psi_{LkgR\gamma} + \Sigma_{k,g,\delta,\gamma}\bar{\psi}_{DLkgL\delta}\,\eta_{DLg}\,m_{DLg\delta\gamma}\psi_{DLkgR\gamma} + \text{c.c.} \tag{4.20}$$

where $m_{Lg\delta\gamma}$ and $m_{DLg\delta\gamma}$ are complex constant matrices, and where δ, γ = 1, … , 4 label generations. The total of fermion lagrangian mass terms is

$$\mathcal{L}^{Higgs}{}_{FermionMasses} = \mathcal{L}_{EWFermionMasses} + \mathcal{L}_{UFermionMasses} + \mathcal{L}_{VFermionMasses} + \mathcal{L}_{GravFermionMasses} + \text{c.c.} \tag{4.21}$$

where $\mathcal{L}_{EW}{}^{Higgs}$ is the contribution of ElectroWeak Higgs Mechanism to the fermion masses (discussed in the following chapter). Using the vacuum expectation value of η in eq. 4.12, and assuming $\eta_{Lg} = \eta_{DLg}$ we find

$$\begin{aligned}\mathcal{L}_{VFermionMasses} = \Sigma_{k,\delta,\gamma}\{&\bar{\psi}_{Lk1L\delta}\rho_1 m_{L1\delta\gamma}\psi_{Lk1R\gamma} + \bar{\psi}_{DLk1L\delta}\rho_1 m_{DL1\delta\gamma}\psi_{DLk1R\gamma} +\\ &+ \bar{\psi}_{Lk2L\delta}\rho_2 m_{L2\delta\gamma}\psi_{Lk2R\gamma} + \bar{\psi}_{DLk2L\delta}\rho_2 m_{DL2\delta\gamma}\psi_{DLk2R\gamma} +\\ &+ \bar{\psi}_{Lk3L\delta}\rho_3 m_{L3\delta\gamma}\psi_{Lk3R\gamma} + \bar{\psi}_{DLk3L\delta}\rho_3 m_{DL3\delta\gamma}\psi_{DLk3R\gamma} +\\ &+ \bar{\psi}_{Lk4L\delta}\rho_4 m_{L4\delta\gamma}\psi_{Lk4R\gamma} + \bar{\psi}_{DLk4L\delta}\rho_4 m_{DL4\delta\gamma}\psi_{DLk4R\gamma}\} + \text{c.c.}\end{aligned} \tag{4.22}$$

where the indices k label species – normal and Dark separately, and the indices δ and γ label *layer* rows and columns. The integers 1, … , 4 label generations.

The mass matrices in eq. 4.21 are complex, constant mass matrices that can be totaled and brought to diagonal form with non-negative values by U(4) matrices. The resulting diagonalized mass matrices are the mass matrices of the physical fermions. See section 4.3 for an example of the procedure.

The fermion masses in the resulting three "upper" layers have terms with similar forms but with different mass values. These values are presumably very large. We expect that they are in the multi-TeV and may extend to tens of TeVs ranges – probably putting most of them out of range of the current CERN LHC.

Due to the weakness of the ultra-weak interaction, but the anticipated large vacuum expectation values of ρ_1, … , ρ_4, the size of the mass cross terms in the Layer group mass matrices of the different layers is problematic. The mass cross terms (mixing) appear likely to be small.

4.7 The Three Generation Alternative

This chapter and the previous chapter developed a formalism based on four generations of fermions, which necessitated a U(4) Generation group and a U(4) Layer group. If there are only three generations of fermions (as presently observed experimentally) then a U(3) Generation group and a U(3) Layer group would follow.

We continue to believe four generations of fermions will be found and thus the U(4) groups are the correct ones.

5. Normal and Dark Matter Dynamics

While fermions are usually identified by name, it will eventually become difficult to use naming conventions when the 192 fermions in four layers are eventually found. We shall introduce a new naming (identification) convention that uniquely specifies individual fermions in the Periodic Table (front cover).

We shall specify a fermion by a triplet of numbers: species s, layer l, and generation g. The identifying triplet is thus (s, l, g).

Species are numbered from 1 through 12: charged lepton, neutral lepton, three up-type quarks, three down-type quarks, Dark charged lepton, Dark neutral lepton, one Dark up-type quark species, and one Dark down-type quark species.

Layers are numbered from 1 through 4 with our layer being layer 1.

Generations are numbered from 1 through 4 from lightest to the heaviest.[49] (e, ν_e, u and d constitute the known part of generation 1.)

For example, the three b quarks are specified by the triplets (6, 1, 2), (7, 1, 2), and (8, 1, 2). The Dark neutral electron neutrino-like fermion of layer 2 corresponds to the triplet (10, 2, 1).

We now turn to consider the ElectroWeak, Generation group and Layer group interactions. Due to the mixing described in prior chapters of generations due to the ElectroWeak and Generation group Higgs Mechanism the interactions below are "dirty" in the sense that they occur between physical fermion and gauge boson particles that intermix generations. The Layer group also intermixes fermion particles between layers to a very small degree due to its ultra-weak interaction – further muddying the interactions. Thus the below diagrams have *very* small admixtures of other states that are not displayed. Fig. 3.1 in section 3.9 shows the Generation and Layer groups fermion mixing.

5.1 ElectroWeak Interactions

The ElectroWeak interactions both for normal and for Dark particles occur within, and between, generations of one layer, and allow transitions between species. The solid lines below represent fermions; the dashed lines represent ElectroWeak gauge bosons. In *general* the form of basic fermion interactions is:

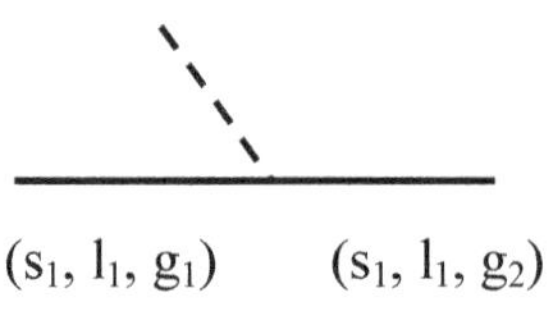

[49] The ordering by mass may not hold in the currently Dark part of the fermion spectrum.

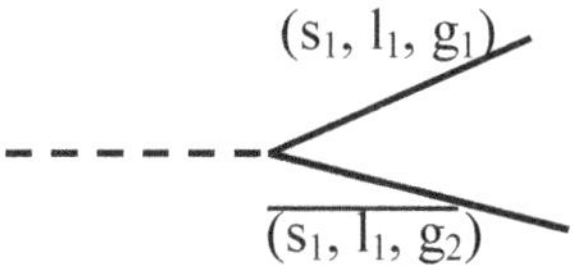

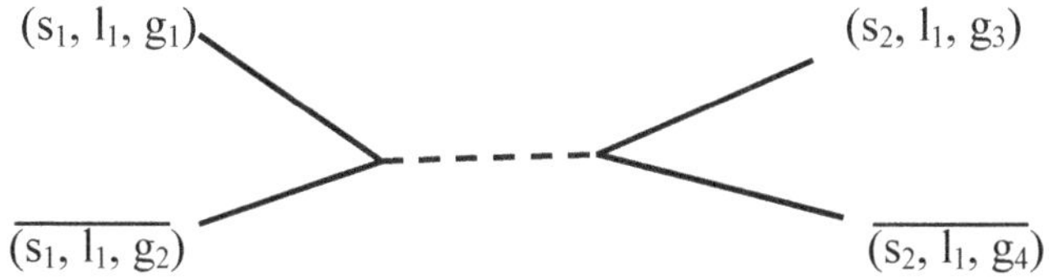

5.2 Generation Group Interactions

The Generation group interactions for both normal and for Dark particles occur within generations of one layer, and within individual species. The 16 gauge field interactions of the Generation group overlap with the four ElectroWeak gauge field interactions since U(4) has an SU(2)⊗U(1) subgroup. Thus they "duplicate" ElectroWeak interactions but "add more." The ElectroWeak and Generation group interactions are distinctively different. The anticipated ultra-weak Generation group coupling constant:

$$g_G = (4\pi\alpha_B)^{1/2} \approx 1.218\ (Gm_H^{\ 2})^{1/2} \tag{16.45}$$

makes their contributions much, much smaller than ElectroWeak contributions to interactions. The solid lines below represent fermions; the dashed lines represent Generation group gauge bosons. In *general* the form of the Generation group interactions is:

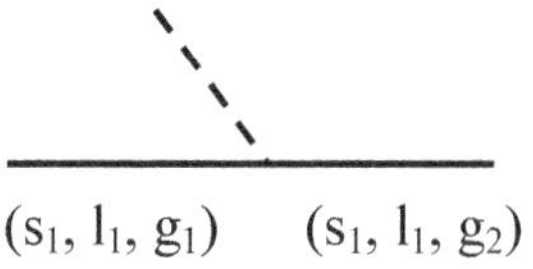

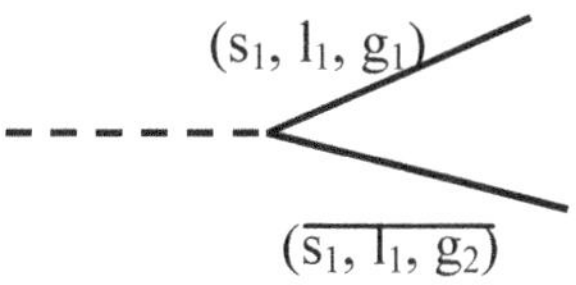

5.3 Layer Group Interactions

The U(4) Layer group interactions for both normal and for Dark particles occur within one layer, and between layers, of each species individually for each of the four generations

individually. It does provide transitions between normal fermions and Dark fermions in our layer (and in other layers) through the four diagonal U(4) Layer interactions: $N\bar{N} \rightarrow \gamma_L \rightarrow D\bar{D}$ where γ_L is a Layer group gauge boson. (Third diagram below)

The ultra-weak Generation group coupling constant g_V, which is also most likely of the order of $(Gm_H^2)^{1/2}$, causes only minimal coupling between the layers – making the layers more or less independent of each other except for rare interactions. Thus it gives a reason for the Darkness of the Dark sector of our layer and of the other layers.

The solid lines below represent fermions; the dashed lines represent Layer group gauge bosons. In *general* the form of the Layer group interactions is:

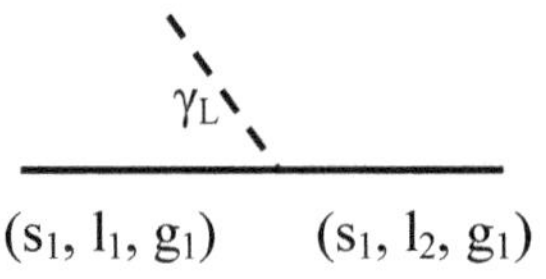

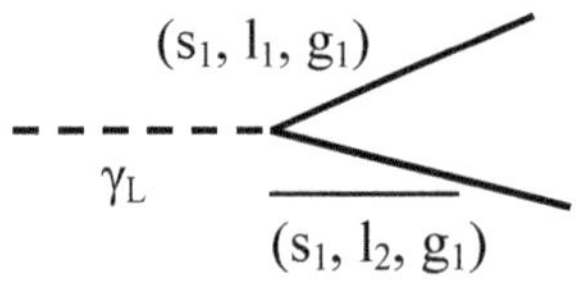

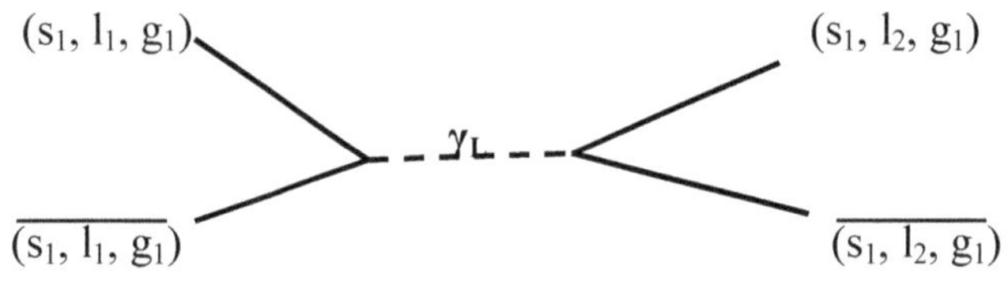

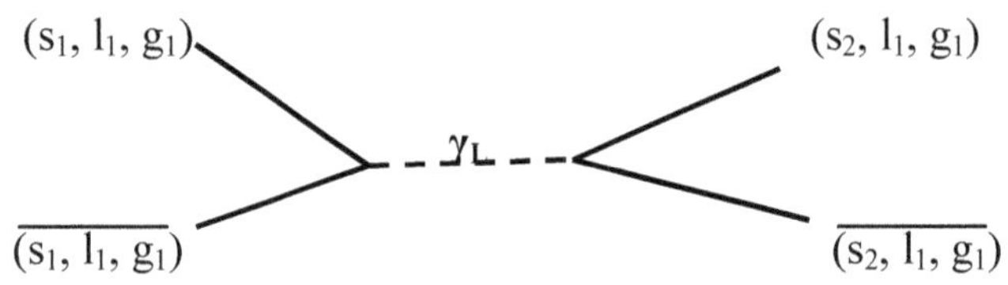

6. The Genesis of Higgs Particle Fields from Complex Gauge Fields

6.1 The Difference between the Strong Gauge Field and the Other Gauge Fields in the Extended Standard Model

In our Extended Standard Model the only gauge field without an associated Higgs particle is the strong interaction gluon gauge field. *We view this exception as a particularly important clue as to the nature of the relation between gauge fields and Higgs particles.*[50]

How does the strong interaction gauge field differ from all other gauge fields in the Extended Standard Model and our Theory of Everything? An examination of the gauge fields dynamic equations (and other lagrangian terms) of our Extended Standard Model reveals that all gauge field dynamic equation kinetic terms *except the strong interaction gauge field* have the form:

$$\partial/\partial x_\mu F^a{}_{\mu\nu} + gf^{abc}A^{b\mu} F^c{}_{\mu\nu} = j^a{}_\nu \tag{6.1}$$

where

$$F^a{}_{\mu\nu} = \partial/\partial x^\nu A^a{}_\mu - \partial/\partial x^\mu A^a{}_\nu + gf^{abc}A^b{}_\mu A^c{}_\nu \tag{6.2}$$

where the coordinates are real-valued, where a, b, c are structure constant indices, where g is a coupling constant, and where $j^a{}_\nu$ is the corresponding current. The gauge field $A^a{}_\mu$ is real for ElectroWeak gauge fields, Generation group gauge fields, and Layer group gauge fields. Thus eqns. 6.1 and 6.2 are real-valued.

The strong interaction gauge field[51] in our Extended Standard Model differs from the other gauge fields by being *necessarily* complex[52] due to the complex 3-space derivatives that appear in the corresponding equations:

$$D^\mu Fc^a{}_{\mu\nu} + gf^{abc}Ac^{b\mu} Fc^c{}_{\mu\nu} = j^a{}_\nu \tag{6.3a}$$

with

$$Fc^a{}_{\mu\nu} = D_\nu Ac^a{}_\mu - D_\mu Ac^a{}_\nu + gf^{abc}Ac^b{}_\mu Ac^c{}_\nu \tag{6.3b}$$

where

$$D_k = \partial/\partial x_r{}^k + i\,\partial/\partial x_i{}^k \tag{6.4}$$
$$D_0 = \partial/\partial x^0$$

for k = 1, 2, 3 where $Ac^a{}_\mu$ is a complexon gauge field. The complex spatial coordinates have the form $x_r{}^k + i\,x_i{}^k$. The time coordinate is real-valued. These equations are eqs. 12.16 and 5.162 of

[50] Most of the material in this chapter appeared in Blaha (2015c).
[51] This field is called a complexon gauge field in Blaha (2015a) and earlier books.
[52] One cannot cleanly separate the real and imaginary parts of its dynamic equations.

Blaha (2015a) for complexon gauge fields,[53] which carry the strong interaction in the Extended Standard Model. Eq. 5.3a requires a complexon gauge field to be complex.

This difference enables us to differentiate the strong gauge field from all other gauge fields in The Extended Standard Model, and thereby to develop a unified formalism for the non-strong gauge fields and their corresponding Higgs particles.

It is the reason that the Strong Interaction gauge fields do not acquire a mass via the Higgs Mechanism. As shown below, the necessary complexity of Strong Interaction gauge fields precludes the generation of Higgs fields from Yang-Mills gauge fields.

6.2 Generation of Higgs Fields From Non-Abelian Gauge Fields

In the prior section we considered the difference between the strong gauge field and the other gauge fields of The Extended Standard Model. Unlike strong gauge fields the other gauge fields (ElectroWeak and so on) could be real or complex. In a manner similar to what we did in the preceding *Physics is Logic* books (and earlier books) we can assume the gauge fields are initially complex, and then transform them to real-valued fields using a phase transformation that introduces scalar fields that we will then take to be Higgs fields.

Since the gauge fields are transformable by Lorentz transformations we can assume each gauge field has a common phase for all its space-time components. Thus we can define a phase transformation for a gauge field $A^{b\mu}$ with

$$A'^{a\mu}(x) = \Phi(x)^a{}_b A^{b\mu}(x) \tag{6.5}$$

where $\Phi(x) = \mathrm{diag}(\exp[i\varphi_1(x)], \exp[i\varphi_2(x)], \ldots, \exp[i\varphi_n(x)])$, and n is the number of symmetry components of $A^{b\mu}$. Inserting $A'^{a\mu}(x)$ in eq. 6.1 we find that eq. 6.1 becomes:

$$\partial/\partial x_\mu F'^a{}_{\mu\nu} + gf^{abc}A'^{b\mu} F'^c{}_{\mu\nu} = j^a{}_\nu \tag{6.6}$$

where

$$F'^a{}_{\mu\nu} = \partial/\partial x^\nu\{\exp[i\varphi_a(x)]A^a{}_\mu\} - \partial/\partial x^\mu\{\exp[i\varphi_a(x)]A^a{}_\nu\} + gf^{abc}\exp[i\varphi_b(x)]\exp[i\varphi_c(x)]A^b{}_\mu A^c{}_\nu \tag{6.7}$$

If we now assume that $\varphi_a(x)$ is small for all a then

$$\exp[i\varphi_a(x)] \simeq 1 + i\varphi_a(x) \tag{6.8}$$

to first order. Substituting in eqs. 6.6 and 6.7, and keeping terms to leading order yields the real part:

---

[53] In The Extended Standard Model we also identify quark species particles as having complex 3-momentum. We call them complexon fermions.

$$\partial/\partial x_\mu F^a_{\ \mu\nu} + gf^{abc}A^{b\mu} F^c_{\ \mu\nu} = j^a_{\ \nu} \tag{6.9}$$

where $F^a_{\ \mu\nu}$ is given by eq. 6.2, and the imaginary part is:

$$\partial/\partial x_\mu F_i{}^a_{\ \mu\nu} + gf^{abc}A^{b\mu} F_i{}^c_{\ \mu\nu} = 0 \tag{6.10}$$

to leading order where

$$F_i{}^a_{\ \mu\nu} = \partial/\partial x^\nu \varphi_a(x)A^a_{\ \mu} - \partial/\partial x^\mu \varphi_a(x)A^a_{\ \nu} \tag{6.11}$$

Substituting eq. 6.11 in eq. 6.10 we find

$$A^a_{\ \nu}\Box\varphi_a(x) - A^a_{\ \mu}\, \partial/\partial x_\mu \partial/\partial x^\nu\, \varphi_a(x) - gf^{abc}A^{b\mu}\, [A^c_{\ \mu}\, \partial/\partial x^\nu\, \varphi_a(x) - A^c_{\ \nu}\, \partial/\partial x^\mu\, \varphi_a(x)] = 0 \tag{6.12}$$

in the Landau gauge, with no sum over a. Eq. 6.12 is a form of Klein-Gordon equation having interaction terms with the gauge field. If the gauge field is weak then only the first two terms are important.

Note that only derivatives of $\varphi_a(x)$ appear in eq. 6.12. Consequently shifts of the $\varphi_a(x)$ field by a constant still yield solutions of eq. 6.12. This feature makes $\varphi_a(x)$ a candidate to be a Higgs particle.

Note also that complexon fields cannot have such a phase change, with a subdivision into real and imaginary dynamic equations, due to the complexity of the spatial coordinates. This difference appears to be the reason why the strong interaction gauge field does not have an associated Higgs particle.

The $\varphi_a(x)$ particles can be made into Higgs particles by adding an appropriate potential:

$$V = A\, \varphi_a^2(x) + B\, \varphi_a^4(x) \tag{6.13}$$

where A and B are constants, or by using our pseudoquantization method. Approximating eq. 6.12 with its first two terms and inserting the potential term we find the Higgs-like equation:

$$A^a_{\ \nu}\Box\varphi_a(x) - A^a_{\ \mu}\, \partial/\partial x_\mu \partial/\partial x^\nu\, \varphi_a(x) + \partial V/\partial\varphi_a = 0 \tag{6.14}$$

$\varphi_a(x)$has a minimum at the minimum of the potential in the corresponding lagrangian.

The second and third terms in eq. 6.14 constitute the interaction. Neglecting these terms we see that eq. 6.14 becomes the free, massless, field Klein-Gordon equation

$$\Box\varphi_a(x) = 0 \tag{6.15}$$

The pairing of Higgs particles with real-valued gauge fields is thus established.[54] The non-existence of a matching Higgs field for the strong interaction is due to the inherently complex nature of the strong interaction (complexon) gauge field in the Extended Standard Model also follows.

To obtain both the vacuum expectation value and the interaction with the quantum part of the pseudoquantum fields we choose to always specify interactions with fermions and gauge fields using $\varphi_a = \varphi_{1a} + \varphi_{2a}$ *since both* φ_{1a} *and* φ_{2a} *satisfy the Klein-Gordan equation eq. 6.15.*

The derivation presented here is analogous to the derivation of Higgs fields in Complex General Relativity – also a gauge theory – in *Physics is Logic Part II.*

One of the remarkable aspects of The Extended Standard Model is its ability to directly prove qualitative properties of elementary particles: four fermion species, Parity violation, the distinction between leptons and quarks, the match of the Standard Models (broken) symmetries with the Reality group consisting of subgroups of U(4), and now the existence of Higgs gauge fields in the ElectroWeak sector but not for the strong interactions. We take these successes to be indicators of the correctness of The Extended Standard Model.

---

[54] Some of the Higgs fields so generated may not have vacuum expectation values and so may only play a role in interactions.

7. Higgs Mechanism for The Theory of Everything Coupling Constants

7.1 Coupling Constants Vacuum Expectation Value Generation

The appearance of just eight fundamental coupling constants in The Extended Standard Model makes them ideal candidates for replacement by eight scalar Higgs particle, vacuum expectation values. *Using vacuum expectation values leads to the remarkable conclusion that all known coupling constants, properly rewritten using our pseudoquantum vacuum expectation value formalism, all have a value of the order of unity – even the gravitational constant G.*[55]

7.1.1 Yang-Mills Coupling Constants

We will first consider the case of a generic Yang-Mills field $A^{b\mu}$ of some symmetry group, a generic fermion field ψ, and a Higgs particle with fields φ_1 and φ_2 (as defined earlier). We will replace its generic coupling constant g with Higgs fields' vacuum expectation values.[56] The initial dynamic equations are

$$\partial/\partial x_\mu F^a{}_{\mu\nu} + gf^{abc}A^{b\mu} F^c{}_{\mu\nu} = j^a{}_\nu \tag{7.1}$$

and

$$[i\gamma^\mu(\partial/\partial x^\mu - igA_\mu) - m]\psi(x) = 0 \tag{7.2}$$

where

$$F^a{}_{\mu\nu} = \partial/\partial x^\nu A^a{}_\mu - \partial/\partial x^\mu A^a{}_\nu + gf^{abc}A^b{}_\mu A^c{}_\nu \tag{7.3}$$

and where a, b, c are structure constant indices, g is the coupling constant, and $j^a{}_\nu$ is the corresponding current.

A gauge transformation has the form

$$gA'_\mu(x) = -i(\partial_\mu\Omega(x))\Omega^{-1}(x) + g\Omega(x)A_\mu(x)\Omega^{-1}(x) \tag{7.4}$$

7.1.2 C-Number Field Coupling Constant

We can replace g with fields in two ways. One way is:

$$\partial/\partial x_\mu F^a{}_{\mu\nu} + m'^{-1}\varphi_1(x)f^{abc}A^{b\mu} F^c{}_{\mu\nu} = j^a{}_\nu \tag{7.5}$$

---

[55] First pointed out in Blaha (2015d).

[56] This approach is conceptually similar to that of Dicke et al for the gravitational constant G. See R. H. Dicke, Phys. Rev. **125**, 2163 (1962) and references therein. See Weinberg (1972) p. 155ff, and Misner et al (1973) p. 1070 for lucid discussions.

and

$$[i\gamma^\mu(\partial/\partial x^\mu - im'^{-1}\varphi_1(x)A_\mu) - m]\psi(x) = 0 \tag{7.6}$$

where

$$F^a_{\mu\nu} = \partial/\partial x^\nu A^a_\mu - \partial/\partial x^\mu A^a_\nu + m'^{-1}\varphi_1(x)f^{abc}A^b_\mu A^c_\nu \tag{7.7}$$

with the corresponding gauge transformation rule:

$$\varphi_1(x)A'_\mu(x) = -im'(\partial_\mu\Omega(x))\Omega^{-1}(x) + \varphi_1(x)\Omega(x)A_\mu(x)\Omega^{-1}(x) \tag{7.8}$$

Using the vacuum state defined by

$$|\Phi, \Pi> = C\exp\{[(2\pi)^3 m/2]^{½}m'g[a_2^\dagger(\mathbf{0},m) + a_2(\mathbf{0},m)]\}|0> \tag{7.9}$$

we see the equations become eqs. 7.1-7.4 since $\Phi = m'g$.
Note m'^{-1} may equal m, or may be the iota Landauer mass[57] or some other value.

Thus we have developed the first Higgs-like mechanism for purely c-number coupling constants.

7.1.3 Q-Number Field Coupling Constant

The other way to reduce coupling constants to vacuum expectation values, which we believe is preferable, is

$$\partial/\partial x_\mu F^a_{\mu\nu} + m'^{-1}(\varphi_1 + \varphi_2)f^{abc}A^{b\mu} F^c_{\mu\nu} = j^a_\nu \tag{7.10}$$

and

$$[i\gamma^\mu(\partial/\partial x^\mu - i\, m'^{-1}(\varphi_1 + \varphi_2)A_\mu) - m]\psi(x) = 0 \tag{7.11}$$

where

$$F^a_{\mu\nu} = \partial/\partial x^\nu A^a_\mu - \partial/\partial x^\mu A^a_\nu + m'^{-1}(\varphi_1 + \varphi_2)f^{abc}A^b_\mu A^c_\nu \tag{7.12}$$

with the corresponding *q-number* gauge transformation rule:

$$(\varphi_1(x) + \varphi_2(x))A'_\mu(x) = -im'(\partial_\mu\Omega(x))\Omega^{-1}(x) + (\varphi_1(x) + \varphi_2(x))\Omega(x)A_\mu(x)\Omega^{-1}(x) \tag{7.13}$$

Using the vacuum state eq. 7.9 we find, for *real-valued* coordinates, eqs. 7.10 – 7.13 become

$$\partial/\partial x_\mu F^a_{\mu\nu} + (g + m'^{-1}\varphi_2)f^{abc}A^{b\mu} F^c_{\mu\nu} = j^a_\nu \tag{7.14}$$

and

$$[i\gamma^\mu(\partial/\partial x^\mu - i(g + m'^{-1}\varphi_2)A_\mu) - m]\psi(x) = 0 \tag{7.15}$$

where

---

[57] The iota mass is a universal mass equal to the Landauer energy of a logical value. See Blaha (2015a).

$$F^{a}_{\mu\nu} = \partial/\partial x^{\nu} A^{a}_{\mu} - \partial/\partial x^{\mu} A^{a}_{\nu} + (g + m'^{-1}\varphi_2) f^{abc} A^{b}_{\mu} A^{c}_{\nu} \quad (7.16)$$

with the corresponding *q-number* gauge transformation rule:

$$(gm' + \varphi_2(x))A'_{\mu}(x) = -i\, m'(\partial_{\mu}\Omega(x))\Omega^{-1}(x) + (gm' + \varphi_2(x))\Omega(x)A_{\mu}(x)\Omega^{-1}(x) \quad (7.17)$$

7.1.4 C-Number Coupling Constants or Q-Number Coupling Constants?

The above two possible methods for reducing coupling constants to Higgsian vacuum expectation values have different experimental implications. In the case of ElectroWeak gauge fields a c-number coupling constant does not introduce a new interaction with a Higgs particle. In the case of ElectroWeak gauge fields a q-number coupling constant, it does introduce a new interaction with a new Higgs particle. The Higgs particle found at CERN LHC may have been produced from an ElectroWeak gauge field with a q-number coupling constant term.

7.1.5 Strong Interaction Case for Real-Valued Coordinates

In the case of the *strong SU(3)* gauge fields the q-number approach would lead to a new interaction of gluons and a Higgs particle corresponding to the field φ_2.

The new dynamic equations for the complexon Yang-Mils field upon replacement of the coupling constant g by a Higgs field using our pseudoQuantization formalism are:

$$D_{\mu} F^{a}_{\mu\nu} + (g + m'^{-1}\varphi_2) f^{abc} A^{b\mu} F^{c}_{\mu\nu} = j^{a}_{\nu} \quad (7.18)$$

and

$$[i\gamma^{\mu}(D_{\mu} - i(g + m'^{-1}\varphi_2)A_{\mu}) - m]\psi(x) = 0 \quad (7.19)$$

where

$$F^{a}_{\mu\nu} = D_{\nu}A^{a}_{\mu} - D_{\mu}A^{a}_{\nu} + (g + m'^{-1}\varphi_2) f^{abc} A^{b}_{\mu} A^{c}_{\nu} \quad (7.20)$$

where all coordinates are *complex-valued* $x = x_r + ix_i$ with derivatives D_{μ} given by eq. 6.4.

The corresponding *q-number* gauge transformation rule:

$$(gm' + \varphi_2(x))A'_{\mu}(x) = -i\, m'(\partial_{\mu}\Omega(x))\Omega^{-1}(x) + (gm' + \varphi_2(x))\Omega(x)A_{\mu}(x)\Omega^{-1}(x) \quad (7.21)$$

Q-number gauge transformations appear in a number of situations. One example of q-number gauge transformations appears in Quantum Electrodynamics.

7.2 Gravitational Coupling Constant Vacuum Expectation Value Generation

The gravitational coupling constant $g_{CG} = \kappa^{-1} = (4\pi G)^{-1/2}$ appears in the gravitational lagrangian density. An example is the case of an interaction with a Dirac particle:

$$\mathcal{L} = g_{CG}^{2} \sqrt{g}\, R/2 + a\bar{\psi}(i\gamma^{\mu}\partial/\partial x^{\mu} - m)\psi \quad (7.22)$$

where a is a coupling constant. and g_{CG} has the dimension of mass. Thus we can introduce a coherent vacuum state

$$|\Phi_G, \Pi_G\rangle = C\exp\{[(2\pi)^3 m/2]^{1/2} g_{CG}[a_2^\dagger(\mathbf{0},m) + a_2(\mathbf{0},m)]\}|0\rangle \tag{7.23}$$

similar to eq. 7.9 that enables us to re-express eq. 7.22 as

$$\mathcal{L} = \varphi_1^2 \sqrt{g}\, R/2 + a\bar{\psi}\,(i\gamma^\mu\partial/\partial x^\mu - m)\psi \tag{7.24}$$

or

$$\mathcal{L} = (\varphi_1 + \varphi_2)^2 \sqrt{g}\, R/2 + c\bar{\psi}\,(i\gamma^\mu\partial/\partial x^\mu - m)\psi \tag{7.25}$$

using the formalism of section 7.4 with the vacuum state eq. 7.23 throughout.[58] Thus we can directly embody the gravitational constant within our formalism.

If we add the pseudoQuantum fields' lagrangian to the lagrangian of eq. 7.24 we obtain:

$$\mathcal{L} = \varphi_1^2 \sqrt{g}\, R/2 + a\bar{\psi}\,(i\gamma^\mu\partial/\partial x^\mu - m)\psi + \partial\varphi_1/\partial x_\mu \partial\varphi_2/\partial x^\mu - m^2\, \varphi_1\varphi_2 \tag{7.26}$$

The dynamic equations for φ_1 and φ_2 are

$$\Box\varphi_2 + m^2\varphi_2 - \varphi_1 \sqrt{g}\, R = 0 \tag{7.27}$$

and

$$\Box\varphi_1 - m^2\varphi_1 = 0$$

In flat space-time, R = 0 and the equations become free field equations. In curved space-time the curvature scalar term becomes a negative mass counter term reminiscent of the corresponding negative term in the Wheeler-DeWitt equation.

The other possible prototype lagrangian

$$\mathcal{L} = (\varphi_1 + \varphi_2)^2 \sqrt{g}\, R/2 + a\bar{\psi}\,(i\gamma^\mu\partial/\partial x^\mu - m)\psi + \partial\varphi_1/\partial x_\mu \partial\varphi_2/\partial x^\mu - m^2\, \varphi_1\varphi_2 \tag{7.28}$$

leads to an interaction between the pseudoQuantum φ_2 field and gravitation:

$$\Box\varphi_2 + m^2\varphi_2 - (\varphi_1 + \varphi_2)\sqrt{g}\, R = 0 \tag{7.29}$$
$$\Box\varphi_1 - m^2\varphi_1 = 0$$

7.3 The Eight Coupling Constants and their Eight PseudoQuantum Field Vacuum Expectation Values

As we mentioned in section 7.1 The Grand Unified Theory of Everything (GUTE) has eight coupling constants:

---

[58] Other variants of these equations are possible such as using the term $g_{CG}\varphi_1\sqrt{g}R/2$ instead of $\varphi_1^2\sqrt{g}R/2$.

- The Strong interaction coupling constant field g_S.
- The ElectroWeak SU(2) coupling constant g_{EW}.
- The ElectroWeak U(1) coupling constant g'_{EW}.
- The Dark ElectroWeak SU(2) coupling constant g_{EWD}.
- The Dark ElectroWeak U(1) coupling constant g'_{EWD}.
- The Layer Group U(4) coupling constant[59] g_V.
- The Generation gauge field U(4) coupling constant g_G.
- The complex gravitational coupling constant $g_{CG} = \kappa^{-1} = (4\pi G)^{-½}$.

Based on the discussions of the previous sections we can define pseudoQuantum fields for these couplings by

- The strong interaction coupling constant vacuum expectation value $\Phi_1 = m_1 g_S$.
- The ElectroWeak SU(2) coupling constant vacuum expectation value $\Phi_2 = m_2 g_{EW}$.
- The ElectroWeak U(1) coupling constant vacuum expectation value $\Phi_3 = m_3 g'_{EW}$.
- The Dark ElectroWeak SU(2) coupling constant vacuum expectation value $\Phi_4 = m_4 g_{EWD}$.
- The Dark ElectroWeak U(1) coupling constant vacuum expectation value $\Phi_5 = m_5 g'_{EWD}$
- The Layer Group U(4) coupling constant vacuum expectation value $\Phi_6 = m_6 g_V$.
- The Generation gauge field U(4) coupling constant vacuum expectation value $\Phi_7 = m_7 g_G$.
- The gravitational coupling constant vacuum expectation value $\Phi_8 = g_{CG} = \kappa^{-1} = (4\pi G)^{-½}$.

The seven masses, m_1 , m_2 , … , m_7 may be equal or they may have different values. It is also possible that all masses may be equal to κ^{-1}, which would yield

- The strong interaction coupling constant vacuum expectation value $\Phi_1 = \kappa^{-1} g_S$.
- The ElectroWeak SU(2) coupling constant vacuum expectation value $\Phi_2 = \kappa^{-1} g_{EW}$.
- The ElectroWeak U(1) coupling constant vacuum expectation value $\Phi_3 = \kappa^{-1} g'_{EW}$.
- The Dark ElectroWeak SU(2) coupling constant vacuum expectation value $\Phi_4 = \kappa^{-1} g_{EWD}$.
- The Dark ElectroWeak U(1) coupling constant vacuum expectation value $\Phi_5 = \kappa^{-1} g'_{EWD}$.
- The Layer Group U(4) coupling constant vacuum expectation value $\Phi_6 = \kappa^{-1} g_V$.
- The Generation gauge field U(4) coupling constant vacuum expectation value $\Phi_7 = \kappa^{-1} g_G$.
- The gravitational coupling constant vacuum expectation value $\Phi_8 = g_{CG} = \kappa^{-1} = (4\pi G)^{-½}$.

Then scaling the above vacuum expectation values by κ^{-1} would give:[60]

- The strong interaction coupling constant[61] vacuum expectation value $\Phi_1' = g_S = 1.22$
- The ElectroWeak SU(2) coupling constant vacuum expectation value $\Phi_2' = g_{EW} = 0.619$.
- The ElectroWeak U(1) coupling constant vacuum expectation value $\Phi_3' = g'_{EW} = 0.347$.
- The Dark ElectroWeak SU(2) coupling constant vacuum expectation value $\Phi_4' = g_{EWD}$. (7.30)

---

[59] This coupling constant appears in Blaha (2016a) and chapter 4.

[60] All coupling constant values are based on data extracted from K. A. Olive et al (Particle Data Group), Chinese Physics **C38**, 090001 (2014).

[61] Based on the running coupling constant value $\alpha_s (M_Z^2) = 0.1193 \pm 0.0016$.

- The Dark ElectroWeak U(1) coupling constant vacuum expectation value $\Phi_5' = g'_{EWD}$.
- The Layer Group U(4) coupling vacuum expectation value $\Phi_6 = g_V$.
- The Generation gauge field U(4) coupling constant vacuum expectation value $\Phi_7' = g_G$.
- The gravitational coupling constant vacuum expectation value $\Phi_8' = 1$.

The *scaled* (known) vacuum expectation values,[62] which are in fact the coupling constants, have a comparable range of values[63] as opposed to the range of values for the unscaled constants which range from the ultra-small gravitational vacuum expectation value to values, perhaps, within a few orders of magnitude of unity.

Given the range of known values above, it appears reasonable to conjecture that the unknown values would also be of the order of unity.

The known coupling constant values in eq. 7.30 are of comparable value, which suggests that our Theory of Everything, at current energies, may be close to the GUT level at which coupling constants are equal.

In the next section we will consider the possibility that the set of scaled vacuum expectation values (the coupling constants) are the elements of a diagonal matrix in a U(8)-like GUTE.

7.4 U(8) GUTE Group

There have been numerous speculations about the ultimate group symmetry that might appear at ultra-high energies. In this chapter we will propose an ultimate U(8) or similar symmetry for The Grand Unified Theory of Everything (GUTE).

In the absence of experimental guidance one must consider the existing observed symmetries (broken and unbroken), and look for signs of the ultimate symmetry. We believe in the likelihood of a U(8)-like GUTE symmetry based on the symmetry of our Theory of Everything symmetry $T_E = SU(3)\otimes SU(2)\otimes U(1)\otimes SU(2)\otimes U(1)\otimes U(4)\otimes U(4)\otimes U(4)$ which combines the observed $SU(3)\otimes SU(2)\otimes U(1)$ symmetries with a U(4) Generation group symmetry to account for the *four* generations of fermions that we anticipate, an additional $SU(2)\otimes U(1)$ symmetry for a Dark ElectroWeak sector for Dark matter that we also anticipate, a U(4) Layer group interaction, and an additional U(4) Reality group symmetry for the Complex Gravity sector.

7.5 Theory of Everything Lagrangian Coupling Constants

We begin with the Theory of Everything lagrangian density with coupling constants explicitly displayed

$$\mathcal{L}_{TE} = \mathcal{L}_{TE}(g_S, g_{EW}, g'_{EW}, g_{EWD}, g'_{EWD}, g_V, g_G, g_{CG}) \qquad (7.31)$$

[62] The closeness of all the values to one is suggestive: The value $\alpha = 1$ (or $e = (4\pi)^{1/2} = 3.54$) was the value found in our calculation in the Johnson, Baker, Willey model of QED (Appendix A). Perhaps a larger calculation along the lines of our paper in massless ElectroWeak theory might yield scaled coupling constant values near unity.

[63] The weakness of the ElectroWeak interactions is primarily due to the large masses of the Z and W vector bosons – not the values of their coupling constants g and g'.

and fields and space-time coordinates not displayed. In terms of vacuum expectation values as discussed earlier we see

$$\mathcal{L}_{TE} = \mathcal{L}_{TE}(\Phi_1/m_1, \Phi_2/m_2, \dots, \Phi_8/m_8) \qquad (7.32)$$

where $g_{CG} = m_8 = \kappa^{-1}$ and where[64]

$$| \Phi_1, \Phi_2, \dots, \Phi_8; \Pi_1, \Pi_2, \dots, \Pi_8> = C \prod_{i=1}^{8} \{\exp[[(2\pi)^3 m_i/2]^{1/2} \Phi_i [a_{i2}^\dagger(\mathbf{0},m_i) + a_{i2}(\mathbf{0},m_i)]]\}|0> \qquad (7.33)$$

Assuming all $m_i = g_{CG} = \kappa^{-1}$ and scaling each mass by g_{CG} we obtain

$$| \Phi_1, \Phi_2, \dots, \Phi_8; \Pi_1, \Pi_2, \dots, \Pi_8> = C \prod_{i=1}^{8} \{\exp[[(2\pi/\kappa)^3/2]^{1/2} \Phi_i' [a_{i2}^\dagger(\mathbf{0}, \kappa^{-1}) + a_{i2}(\mathbf{0}, \kappa^{-1})]]\}|0> \qquad (7.34)$$

in conformity with ea. 7.30. Then eq. 7.32 can be written as

$$\mathcal{L}_{TE} = \mathcal{L}_{TE}(\Phi_1', \Phi_2', \dots, \Phi_8') \qquad (7.35)$$

which is in agreement with eq. 7.30.

Setting $m_i = g_{CG} = \kappa^{-1}$ = the Planck mass, simplifies the above expressions and *supports the unification of all interactions*. However having particles of such large mass makes them undetectable by accelerators. It also seems too large from the viewpoint of physical intuition. Consequently, eqs. 7.32 and 7.33 may be the correct expressions. We finally note

$$\varphi_{1i}| \Phi_1, \Phi_2, \dots, \Phi_8; \Pi_1, \Pi_2, \dots, \Pi_8> = \Phi_{1i}| \Phi_1, \Phi_2, \dots, \Phi_8; \Pi_1, \Pi_2, \dots, \Pi_8> \qquad (7.36)$$

7.6 Coupling Constant Symmetry

In the GUTE limit, if it exists, as the closeness of the scaled, known coupling constant values in eq. 7.30 suggest, the scaled coupling constants (vacuum expectation values) in The Theory of Everything lagrangian have an S_8 symmetry. The lagrangian is symmetric under any of the 8! permutations of the eight vacuum expectation values.

The appearance of S_8 symmetry for the scaled coupling constants is interesting in the light of numerous attempts to understand mass and mixing matrices using various symmetry groups such as S_4.

7.7 Big Bang Vacuum

At the origin of the universe – the Big Bang – there was a vacuum state in principle. In our earlier books[65] we showed that the universe existed in an ultra-small, but finite, region for

---

[64] The "vacuum" state |0> also has factors for the vacuum expectation values used for fields that give masses to fermions and vector bosons as described in Blaha (2015b).

[65] Blaha (2015a) and Blaha (2004).

an infinitesimal time before it began an explosive inflationary expansion to become the familiar universe. In this time period there were no infinities – a finite temperature and so on.

Thus it is reasonable to assume one of two possibilities for the eight coupling constants: 1) they have remained unchanged since the beginning, or 2) they have changed with time.

In this section we note, that if our scaling with the Planck mass κ^{-1} in preceding discussions is correct, then it is reasonable to assume that the vacuum state in the beginning is eq. 7.34 with |0> including factors setting fermion and vector boson masses as described in Blaha (2015b).

7.8 Evolving Coupling Constants

If coupling constants evolved over time, as some think, then we would need a more complex vacuum state. At present most cosmological data supports unchanged coupling constants in time and spatially. Nevertheless this possibility is not ruled out.

7.9 Space-Time Dependent Coupling Constants

It is possible that the Theory of Everything coupling constants evolve with time and may also be spatially varying – different constants in different parts of the universe. Presently there is no decisive evidence for either possibility. In this chapter we will describe the mechanism to support either or both possibilities.

Consider a classical field (time and spatially varying):

$$\Phi(\mathbf{x}, t) = \int d^3k \, [\alpha(k) f_k(x) + \alpha^*(k) f_k^*(x)] \tag{7.37}$$

If we define the coherent vacuum state

$$|\alpha\rangle = C \exp\left\{\int d^3k \, [\alpha(k) a_2^\dagger(k) + \alpha^*(k) a_2(k)]\right\} |0\rangle \tag{7.38}$$

then

$$\begin{aligned} \varphi_1(x)|\Phi, \Pi\rangle &= \Phi(x)|\Phi, \Pi\rangle \\ \pi_1(x)|\Phi, \Pi\rangle &= \Pi(x)|\Phi, \Pi\rangle \end{aligned} \tag{7.38}$$

where

$$\varphi_i(\mathbf{x}, t) = \int d^3k \, [a_i(k) f_k(x) + a_i^\dagger(k) f_k^*(x)] \tag{7.39}$$

for i = 1, 2 and

$$f_k(x) = e^{-ik\cdot x} / (2\omega_k (2\pi)^3)^{1/2}$$

with $\omega_k = |\mathbf{k}|$.

Eq. 7.38 contains the coherent state $|\alpha>$ for a time and spatially varying vacuum expectation value (classical) field. The above equations can be generalized to the case of the eight coupling constant vacuum expectation values:[66]

$$| \Phi_1, \Phi_2, \dots , \Phi_8; \Pi_1, \Pi_2, \dots , \Pi_8> = C \prod_{i=1}^{8} \exp\{\int d^3k\, [\alpha_i(k)a_{2i}^\dagger(k) + \alpha_i^*(k)a_{2i}(k)]\}|0> \quad (7.40)$$

Then all eight coupling constant vacuum expectation values are space-time dependent:

$$\varphi_{1i}(x) \,| \Phi_1, \Phi_2, \dots , \Phi_8; \Pi_1, \Pi_2, \dots , \Pi_8> = \Phi_i(x) \,| \Phi_1, \Phi_2, \dots , \Phi_8; \Pi_1, \Pi_2, \dots , \Pi_8> \quad (7.41)$$

and the Theory of Everything lagrangian

$$\mathcal{L}_{TE} = \mathcal{L}_{TE}(g_S, g_{EW}, g'_{EW}, g_{EWD}, g'_{EWD}, g_{EWD}'', g_G, g_{CG}) \quad (7.42)$$

becomes

$$\mathcal{L}_{TE} = \mathcal{L}_{TE}(\Phi_1(x), \Phi_2(x), \dots , \Phi_8(x)) \quad (7.43)$$

for matrix elements between the vacuum and its conjugate,

Thus our formalism can accommodate space-time varying coupling constants should they be found in the Cosmos.

7.10 A Theory of Everything Lagrangian Without Any Constants

The preceding chapters put coupling constants within the same framework as particle masses completing the process of eliminating constants from The Theory of Everything lagrangian. Instead the vacuum contains the values of all coupling constants and particle masses. In one sense this new formulation is a tradeoff. The values of all constants are shifted to the vacuum. However the shift has some advantages technically. One advantage is the ability to have space-time dependent coupling constants. It would also be straightforward to make masses, and mixing angles, space-time dependent. The possible space-time dependence of coupling constants and particle masses has been an active area of experimental interest for many years although cosmological data seems to indicate these quantities have not changed significantly since the universe began.

The question of changes in lagrangian constant physical values is of great philosophical importance since it appears that the existence of life, as we know it, depends sensitively on their values. This dependence has been embodied in the Anthropomorphic Principle and studied by a number of physicists and philosophers.

Since our formulation allows space-time varying physical constants the question of the Anthropomorphic Principle attains new importance. As we saw in the case of the theory of Black Holes, which was a theory without evidence for over forty years before Black Holes were

---

[66] The "vacuum" state |0> in eq. 5.34 also has factors for the vacuum expectation values used for fields that give masses to fermions and vector bosons as described in Blaha (2015b).

discovered, Nature seems to provide phenomena that have been shown to be theoretically possible. Many other cases of this sort have also occurred – the most recent example at the time of this writing is Weyl fermions.

Lastly, our approach opens the possibility of a study of all the many constants in The Theory of Everything lagrangian *on the same footing* rather than in the piecemeal fashion used up to the present. It replaces the scattered hodge-podge of constants in the lagrangian with a centralized location for all constants in the vacuum state permitting a direct study of their interconnection. *The study of the vacuum now becomes of central importance.*

7.11 The Form is Determined But Not the Constants

The derivation of The Extended Standard Model in the Blaha (2015a) and earlier work was based on Asynchronous Logic (to support physical processes spread in space and time); on complex space-time coordinates, the Complex Lorentz group and complex General Coordinate transformations; the Reality group to map complex coordinates to the real-valued coordinates that we observe; the Generation group that yields the four fermion generations and particle number interactions such as the baryon number and lepton number interactions, the Layer group to generate fermion layers and normal-Dark particle interactions, and the Reality group for complex general coordinate transformations.

This firm basis in fundamental considerations enables us to forge a path to The Extended Standard Model, which included the known features of The Standard Model. We thus were able to avoid the many possible variants and extensions of The Standard Model that have been considered in the Physics literature over the past thirty years.

Two remarkable features of The Extended Standard Model derivation were:

1. A precise fixing of the form of the Extended Standard Model and the Theory of Everything.

2. The absence of any constraints on the values of its coupling constants or masses.

Particle masses were fixed by either the original Higgs Mechanism or by our new mechanism that was based on an extension of Quantum Field Theory to include classical fields such as the vacuum expectation values that cropped up in the original Higgs Mechanism and were handled "by hand." (See Blaha (2015c).)

Thus, up to this point, we have a Theory of Everything (known) except for a basis for the values of the coupling constants that appear in the theory. The coupling constants have a wide range of values. A fundamental basis for their values has been wanting. We will suggest a mechanism for the determination of their values and a (badly broken) U(8) symmetry for The Theory of Everything and the eight coupling constants in The Extended Standard Model.

7.12 Coupling Constants of The Extended Standard Model

The Theory of Everything includes The Extended Standard Model and our theory of Complex General Relativity (which is well-approximated by Einstein's Theory of General Relativity.)

$SU(3)\otimes SU(2)\otimes U(1)\otimes SU(2)\otimes U(1)\otimes U(4)\otimes U(4)\otimes U(4)$ is the symmetry group of The Extended Standard Model. Our Theory of Complex General Relativity has a U(4) Reality group that yields a set of Higgs-like fields that contributes to fermion masses. Each of the subgroups of the Extended Standard Model has an associated coupling constant. Complex General Relativity also has an associated coupling constant G.

Thus The Theory of Everything has eight coupling constants:

1. The strong interaction coupling constant which we denote g_S.
2. The ElectroWeak SU(2) coupling constant g_{EW}.
3. The ElectroWeak U(1) coupling constant g'_{EW}.
4. The Dark ElectroWeak SU(2) coupling constant g_{EWD}.
5. The Dark ElectroWeak U(1) coupling constant g'_{EWD}.
6. The Layer group U(4) coupling constant g_V.
7. The Generation gauge field U(4) coupling constant g_G.[67]
8. The gravitational coupling constant $g_G = \kappa^{-1} = (4\pi G)^{-½}$.[68]

7.13 How Can Coupling Constants be Determined?

The renormalized coupling constants of The Theory of Everything can be determined experimentally. However a theoretical determination is lacking. This gap in our understanding suggests that there is a major aspect of fundamental physics that is not understood. The fact that we can determine the form – but not the values of coupling constants – so directly from basic principles suggests that new basic principles are needed to complete The Extended Standard Model. A similar comment applies to fermion and boson masses – both our mechanism,[69] and the vanilla Higgs Mechanism, arbitrarily fixes particle masses. (Attempts to relate particle masses using various symmetries beg the question. As Isidore Rabi (Columbia) once said in a different context, "Who ordered them." Proposed symmetries are typically "pulled out of a hat.")

The *one* meaningful attempt to determine a coupling constant in a non-trivial 4-dimensional quantum field theory was that of Johnson, Baker and Willey[70] in a 4-dimensional model – massless Quantum Electrodynamics. They developed the theory to the point where if one function, that they called the eigenvalue function, had a zero at the value of the fine structure constant $\alpha \approx 1/137$ then the theory would have no infinities. Adler then showed that the eigenvalue constant zero must be an essential singularity, IF it had a zero. This author then

---

[67] See Blaha (2015a) p. 194 and also Blaha (2015b).
[68] Blaha (2015c).
[69] Blaha (2015c).
[70] M. Baker and K. Johnson, Phys. Rev. **D8**, 1110 (1973) and references therein.

developed an approximate solution for the eigenvalue function in perhaps the most comprehensive 4-dimensional quantum field theory calculation to all orders in α. The approximate calculation agreed with known exact results to 6th order in e.[71] The author found a zero at α = 1. The zero was not an essential singularity.[72]

While the Johnson, Baker, Willey model QED was not successful in finding α its method illustrates one possible approach to determining the coupling constants of The Extended Standard Model. It might be possible to use a consistency condition(s) to fix coupling constant values. Since The Extended Standard Model does not have infinities, the motivation of Johnson, Baker and Willey is absent. A fundamental set of consistency conditions is not apparent and so this approach is not currently viable.

What other approaches are possible? There is an anthropomorphic approach which posits the necessity of certain ranges of some coupling constants for human life, and life in general, to exist. We are not comfortable with this approach since it seems to "beg the question." The input is equivalent to the output – mitigating its character as fundamental.

One could also study the set of coupling constants in a 8-dimensional space looking for a set of values with especially beneficial results – perhaps using Leibniz's Minimax principle: a universe with the "broadest" set of physical phenomena and the least assumptions. The ambiguity of this approach, at the present time, should be apparent to the reader.

Given these considerations we have chosen to pursue a less ambitious approach: to specify the coupling constants as vacuum expectation values of a set of new Higgs-like scalar fields. This approach conceptually parallels the determination of particle masses as vacuum expectation values of scalar Higgs fields.

After reducing coupling constants to vacuum expectation values we consider the possibility that the vacuum state at the Big Bang point determined the coupling constants. We also consider the possibility that coupling constants evolve slowly with time and/or may vary in differing spatial locations.

---

[71] Equation 1 in our paper S. Blaha, Phys. Rev. **D9**, 2246 (1974) in Appendix A.

[72] In retrospect the realization that QED was part of ElectroWeak theory made success doubtful.

8. Higgs Non-Abelian Vector Boson Pseudoquantum Field Theory

Non-abelian Yang-Mills (gauge) Quantum Field Theory are well known. It is possible to formulate a non-Abelian pseudoquantum field theory similar to our pseudoquantum scalar field theory described earlier. Pseudoquantum non-abelian field theory[73] is based on the expansion of conventional non-abelian quantum field theory features such as the covariant derivative to include an affine connection term.

8.1 Affine Connection for Non-Abelian Pseudoquantum Field Theory

The affine connection is most often viewed as part of the derivation of the curvature tensor in General Relativity. It is typically derived from manipulations of the metric $g_{\mu\nu}$. However, the affine connection can also be viewed as a set of independent fields that become related to the metric via dynamic equations.

Some years ago A. Einstein and H. Weyl[74] pointed out that the metric and the affine connection should be treated as independent quantities and subject to independent arbitrary infinitesimal variations:

> "In contrast to Einstein's original "metric" conception in terms of the $g_{\nu\mu}$ there was later developed, by Eddington, by Einstein himself, and recently by Schrödinger, an affine field theory operating with the components $\Gamma^{\sigma}{}_{\nu\mu}$ of an affine connection. But in 1925 Einstein also advocated a "mixed" formulation by means of a lagrangian in which both the $g_{\nu\mu}$ and the $\Gamma^{\sigma}{}_{\nu\mu}$ are taken as basic field quantities and submitted to independent arbitrary infinitesimal variations.[75] In certain respects this seems to be the most natural procedure."

This approach leads to a new view of non-abelian pseudoquantum field theory that we will consider in this chapter.

8.2 Generalization of the Yang-Mills Non-Abelian Covariant Derivative

The usual covariant derivative for a non-Abelian gauge theory is

$$D_\nu = \partial_\nu + igA^1{}_\nu \times \qquad (8.1)$$

[73] This formulation was first presented in S. Blaha, Phys. Rev. D**11**, 2921 (1975). See Appendix 8-A.
[74] H. Weyl, Phys. Rev. **77**, 699 (1950).
[75] A. Einstein, Sitzungsber., Preuss. Akad. Der Wissensch. (1925), p. 414.

where $A^1{}_\nu$ is a non-Abelian gauge field to which we provisionally add the index "1" and where g is a constant.

We now introduce an additional term – an affine connection $A^{2\sigma}{}_{\nu\mu}$ with the index "2" added – that we treat as independent variables subject to independent variations. We apply this derivative to a vector V_μ:

$$D_\nu V_\mu = (\partial_\nu + igA^1{}_\nu\times)V_\mu + igA^{2\sigma}{}_{\nu\mu}\times V_\sigma \tag{8.2}$$

We now define the Yang-Mills affine connection in terms of a second gauge field:

$$A^{2\sigma}{}_{\nu\mu} = g^\sigma{}_\mu A^2{}_\nu \tag{8.3}$$

Under a local gauge transformation C(x) we achieve covariance by defining the fields' gauge transformations by

$$\begin{aligned} A^{1\mu}(x) &\to C(x)A^{1\mu}(x)C^{-1}(x) - i\,C(x)\partial^\mu C^{-1}(x)/g \\ A^{2\mu}(x) &\to C(x)A^{2\mu}(x)C^{-1}(x) \end{aligned} \tag{8.4}$$

Defining the field strengths as

$$\begin{aligned} F^1{}_{\mu\nu} &= \partial_\mu A^1{}_\nu - \partial_\nu A^1{}_\mu - i\,gA^1{}_\mu\times A^1{}_\nu \\ F^2{}_{\mu\nu} &= \partial_\mu A^2{}_\nu - \partial_\nu A^2{}_\mu + igA^1{}_\mu\times A^2{}_\nu - igA^1{}_\nu\times A^2{}_\mu \end{aligned} \tag{8.5}$$

The lagrangian is

$$\mathcal{L} = \tfrac{1}{2}\, F^1{}_{\mu\nu}\cdot F^{2\mu\nu} \tag{8.6}$$

The equations of motion[76] are

$$\partial/\partial x^\mu\, F^{1a}{}_{\mu\nu} + gf^{abc}A^{1b\mu}\, F^{1c}{}_{\mu\nu} = 0 \tag{8.7}$$

$$\partial/\partial x^\mu\, F^{2a}{}_{\mu\nu} + gf^{abc}A^{1b\mu}\, F^{2c}{}_{\mu\nu} + gf^{abc}A^{2b\mu}\, F^{1c}{}_{\mu\nu} = j^a{}_\nu \tag{8.8}$$

Since the present development of this non-Abelian theory is similar to that presented in Appendix 8-A with λ = 0, we can use the definition of the canonically conjugate momenta in eqs. 50 and 51, together with the equal time commutation relations for the independent dynamical variables in eqs. 65 and 66, to establish a pseudoquantum formalism in which one field can be taken to be a "classical" field $A^{2b\mu}$ and the other field $A^{1b\mu}$ a quantum field realizing that higher order terms in the lagrangian will add quantum pieces to the "classical" field.

Sections 2.2 – 2.4 describe the scalar field case which can be transposed to the non-Abelian case here by interchanging field indexes 1 and 2, and by adding the group indices.

Consequently

---

[76] See S. Blaha, Phys. Rev. D**11**, 2921 (1975) in Apendix 8-A.

$$\bar{\psi}\gamma_\mu T^b(A^{1b\mu} + A^{2b\mu})\psi \rightarrow \bar{\psi}\,\gamma_\mu T^b(A^{1b\mu} + \Phi^{2b\mu})\psi \qquad (8.9)$$

under a Higgs Mechanism where $\Phi^{2b\mu}$ is a constant vector. *Thus we have a formalism that enables a form of Higgs Mechanism for non-Abelian fields.*

"Almost" classical non-Abelian gauge fields, such as $A^{2b\mu}$, may be relevant in the case of quark-gluon plasmas, which have been created at Brookhaven and CERN recently.

Chapter 9 considers the case of a spin 1 pseudoquantum field theory, which contains some details relevant to the present discussion.

Appendix 8-A. Pseudoquantum Non-Abelian Field Theory Paper

This refereed paper is S. Blaha, Phys. Rev. D**11**, 2921 (1975).

Second-quantized non-Abelian field theory for hadrons with quark confinement and scaling deep-inelastic structure functions*

Stephen Blaha

Laboratory of Nuclear Studies, Cornell University, Ithaca, New York 14853

(Received 30 December 1974)

A four-dimensional second-quantized field theory with quarks bound by "colored" non-Abelian gluons is described which has the following properties: (1) the only physical particles are color singlets composed solely of quarks, (2) the deep-inelastic structure functions have Bjorken scaling, (3) gluon loops and Faddeev-Popov ghost loops are identically zero in any gauge, (4) Regge trajectories are apparently linear on a Chew-Frautschi plot, and (5) constituent motion within hadrons can be nonrelativistic.

I. INTRODUCTION

After a period of some skepticism the possibility that hadronic interactions might be understood within the framework of quantum field theory is again being seriously considered.[1] This is partly the result of the psychological climate created by the apparently successful unification of weak and electromagnetic interactions in a renormalizable field theory and partly the result of a greater appreciation of the variety of phenomena which can occur in field theories.

In this article we shall describe a field-theoretic model of hadron binding which has two major features: (1) Hadrons only occur as quark-antiquark or three-quark bound states, and (2) quarks behave as quasifree particles within hadrons. We assume that the suggestions of an internal symmetry called color[2] are correct and that the strong interaction consists of the exchange of colored Yang-Mills gluons. The nature of the interaction allows only color singlet states to occur in the gauge-invariant physical particle spectrum and consequently the first feature will be realized by choosing the color group to be SU(3). Since the (Schwinger) mechanism which produces this result is an infrared phenomenon, the second feature is not precluded and the model is essentially free in the ultraviolet region of the quark sector.

Our model is a non-Abelian version of a recently investigated Abelian field theory which had quark confinement and scaling electroproduction structure functions.[3] In that theory the free propagator of the massless gluon field embodying the quark-quark interaction was proportional to

$$\lambda^2/k^4, \tag{1}$$

where λ is a constant with the dimensions of mass and k is the gluon four-momentum. As a result the Schwinger mechanism[4] manifestly occurred, and it was shown that any charged particle was totally screened by vacuum polarization effects. In addition, explicit calculations of the deep-inelastic electroproduction structure functions in perturbation theory were in agreement with Bjorken scaling with corrections of $O(q^{-4})$, where q is the virtual photon four-momentum. These features of the Abelian model will also be shown to be true in the non-Abelian version. In addition, we shall argue that the quarks can be nonrelativistic within hadrons and that the spectrum of states has linearly rising Regge trajectories.

In spite of these salutary properties an interaction of the form of Eq. (1) could be questioned because of well-known[5] indefinite-metric difficulties which result in the violation of unitarity. While an optimist may hope that the nonappearance of colored gluons in asymptotic (color singlet) states might eliminate unitarity problems it is almost certain that the approximation techniques which will necessarily be used to find the bound states will lead to the occurrence of negative-metric states. Whether these states are "real" or artifacts of the approximation will not be clear. In view of this we suggested[3] that the gluon propagator be taken in principal value rather than as a Feynman propagator:

$$\mathrm{P}\frac{\lambda^2}{k^4}=\frac{\lambda^2}{2}\left[\frac{1}{(k^2+i\epsilon)^2}+\frac{1}{(k^2-i\epsilon)^2}\right]. \tag{2}$$

As a result unitarity is maintained order by order in perturbation theory. Gluons do not appear in asymptotic states. All components of the vector-gluon propagator are "Coulombized" and the gluon field reduced to the embodiment of a direct quark interaction. There are a number of other decided advantages to principal-value propagators in the present context: (1) no color singlet states composed solely of gluons, (2) the elimination of substantial infrared divergences, (3) the suppression of corrections to Bjorken scaling in the electroproduction structure functions by a factor of q^2

vis-à-vis the corresponding Feynman-propagator result which sets the stage for precocious scaling, and (4) the elimination of closed loops of vector gluons and thus the elimination of Faddeev-Popov ghost loops.

In Sec. II we give a brief recapitulation of the Abelian model. In Sec. III we describe the canonical properties of the non-Abelian model. In Sec. IV we describe the qualitative features of the model and describe an approximation technique which appears to be naturally adopted to "solving" the theory. We shall restrict our discussion to the color binding interaction and defer the introduction of other interactions to a later work. The properties of the bound states in the non-Abelian model are currently under study and will be the subject of the next report.

II. ABELIAN MODEL

The possibility that the physical particle spectrum of a field theory consisted only of neutral states and did not include states of charged fields was first investigated in massless two-dimensional quantum electrodynamics.[6] In that case the absence of the "electron" from the gauge-invariant physical particle spectrum was directly related to the acquisition of a mass by the photon via the Schwinger mechanism. The Schwinger mechanism was manifest in the lowest-order contribution to the vacuum polarization (Fig. 1), and, taking account of the dimensionality of the coupling constant, $e\sim$ mass, could almost be considered a consequence of dimensional analysis. These vacuum polarization effects led to the total screening of the "electronic" charge, and, as a result, the "electron" was removed from the gauge-invariant physical particle spectrum. Our Abelian and non-Abelian models will display a similar pattern of events.

The Lagrangian of the Abelian model contains two gluon fields, $A^1_\mu(x)$ and $A^2_\mu(x)$, and the quark field $\psi(x)$:

$$\mathcal{L}=-\tfrac{1}{2}F^1_{\mu\nu}F^2_{\mu\nu}-\tfrac{1}{2}\lambda^2A^2_\mu A^2_\mu+\bar{\psi}(i\nabla\!\!\!/-gA\!\!\!/^1-m)\psi\,, \tag{3}$$

where for typographic convenience we denote the inner product of four vectors, $a\cdot b=a_\mu b_\mu=a_0b_0-\vec{a}\cdot\vec{b}$ throughout, λ is a constant with the dimensions of mass, g is dimensionless, and $F^i_{\mu\nu}=\partial_\nu A^i_\mu-\partial_\mu A^i_\nu$.

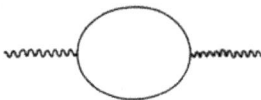

FIG. 1. A vacuum polarization diagram.

Following the canonical procedure we find the equations of motion:

$$\partial_\mu F^1_{\mu\nu}+\lambda^2A^2_\nu=0\,, \tag{4}$$

$$\partial_\mu F^2_{\mu\nu}+gJ_\nu=0\,, \tag{5}$$

$$(i\nabla\!\!\!/-gA\!\!\!/^1-m)\psi=0\,, \tag{6}$$

and nonzero equal-time commutation relations [in the Coulomb gauge $\vec{\nabla}\cdot\vec{A}^1=0$; note $\partial_\mu A^2_\mu=0$ by Eq. (4)]

$$[F^1_{0i}(x),A^2_j(y)]=i\Delta^{tr}_{ij}(x-y)\,, \tag{7}$$

$$[F^2_{0i}(x),A^1_j(y)]=i\Delta^{tr}_{ij}(x-y)\,, \tag{8}$$

with $i,j=1,2,3$ and

$$\Delta^{tr}_{ij}(x-y)=\int\frac{d^3k}{(2\pi)^3}e^{i\vec{k}\cdot(\vec{x}-\vec{y})}\left(\delta_{ij}-\frac{k_ik_j}{|\vec{k}|^2}\right). \tag{9}$$

It is clear from the equations of motion, Eqs. (4) and (5), tnat A^2_μ may be eliminated to obtain

$$\Box\partial_\mu F^1_{\mu\nu}+g\lambda^2J_\nu=0\,. \tag{10}$$

The form of the quark-gluon interaction and Eq. (10) show that only the Green's function of A^1_μ is relevant to quark-quark scattering. The perturbation theory rules of QED may be used if the photon propagator is replaced with the gluon propagator for A^1_μ:

$$iG^{11}_{\mu\nu}(k)=\frac{i\lambda^2(g_{\mu\nu}-\chi k_\mu k_\nu/k^2)}{k^4}\,, \tag{11}$$

where χ is constant, determined by the gauge choice.

In Ref. 3 we showed that choosing $G^{11}_{\mu\nu}$ to be a principal-value propagator allowed us to develop a perturbation theory which was unitary order by order:

$$G^{11}_{\mu\nu}(k^2)\equiv\tfrac{1}{2}[G^{11}_{\mu\nu}(k^2+i\epsilon)+G^{11}_{\mu\nu}(k^2-i\epsilon)]\,. \tag{12}$$

In addition, the equivalent of the Nambu representation of a Feynman diagram was given and some features of the perturbation theory discussed. Of particular interest was a calculation of the deep-inelastic electroproduction structure functions which scaled in the Bjorken limit. Leading corrections to scaling were of $O(q^{-4})$ as $q^2\to\infty$ with q being the virtual photon four-momentum, and were given by the diagrams of Fig. 2(b), 2(c), and 2(d). This is to be contrasted with the logarithmic deviations from scaling found in pseudoscalar or vector meson models previously studied.[7]

The Schwinger mechanism manifestly occurred in low orders of perturbation theory. As a result quarks (and all charged objects) are removed from the gauge-invariant spectrum of physical states. The total screening of charge can be seen from the following argument.[3] Consider a spatially bounded system of charge density ρ. The total charge is

$$Q = \int d^3x\, \rho(x) \tag{13}$$

$$= \frac{-1}{g\lambda^2} \int d^3x\, \Box \nabla^2 A_0^1 \tag{14}$$

using the equations of motion in the Coulomb gauge. By Gauss's law

$$Q = \frac{-1}{g\lambda^2} \int d\vec{S} \cdot \vec{\nabla} \Box A_0^1 . \tag{15}$$

From the definition of a Green's function, we have

$$A_0^1(x) = \int d^4y\, G_{00}^{11}(x-y)\rho(y) \tag{16}$$

in the Coulomb gauge. If, for simplicity, we choose ρ to describe a static point quark charge and use the free gluon propagator [Eq. (11)], then $Q \neq 0$. However, if we take account of the effect of vacuum polarization processes (the Schwinger mechanism) we find $A_0^1(x)$ is a monotonically decreasing function of $|\vec{x}|$ for large $|\vec{x}|$ and consequently $Q = 0$ in the limit where the integration surface is taken to infinity in Eq. (15). Thus the spectrum of physical states does not include states of nonzero charge. In the next section we shall show that the proof of quark confinement is essentially the same in the non-Abelian model.

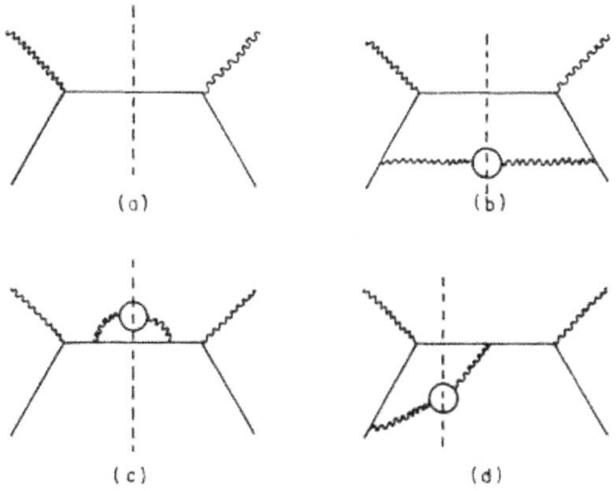

FIG. 2. Lowest-order diagrams contributing to the inelastic electroproduction structure functions. The dashed lines indicate the only contributions to the electroproduction structure functions of the absorptive part of the forward virtual Compton scattering diagram. External "wiggly" lines represent photons while internal "wiggly" lines represent gluons.

III. NON-ABELIAN MODEL

The non-Abelian model for the color sector of hadronic interactions is a direct generalization of the model of the last section.[8] There are two colored Yang-Mills fields, $A_{\mu a}^1(x)$ and $A_{\mu a}^2(x)$, which when regarded as vectors in the adjoint representation of the color group are denoted $\underline{A}_\mu^1$ and $\underline{A}_\mu^2$. The Lagrangian is

$$\begin{aligned}\mathfrak{L} = {} & \tfrac{1}{2}\underline{F}_{\mu\nu}^1 \cdot \underline{F}_{\mu\nu}^2 - \tfrac{1}{2}\underline{F}_{\mu\nu}^2 \cdot (\partial_\mu \underline{A}_\nu^1 - \partial_\nu \underline{A}_\mu^1 + g\underline{A}_\mu^1 \times \underline{A}_\nu^1) \\ & - \tfrac{1}{2}\underline{F}_{\mu\nu}^1 \cdot (\partial_\mu \underline{A}_\nu^2 - \partial_\nu \underline{A}_\mu^2 + g\underline{A}_\mu^1 \times \underline{A}_\nu^2 - g\underline{A}_\nu^1 \times \underline{A}_\mu^2) \\ & - \tfrac{1}{2}\lambda^2 \underline{A}_\mu^2 \cdot \underline{A}_\mu^2 + \bar{\psi}(i\nabla + g A^1 - m)\psi \end{aligned} \tag{17}$$

$$= \mathfrak{L}_0 + \bar{\psi}(i\nabla + g \not{A}^1 - m)\psi , \tag{18}$$

with ψ being the quark field.

It is invariant under the local gauge transformation

$$\psi' = S^{-1}\psi , \tag{19}$$

$$A_\mu^{1\prime} = S^{-1}A_\mu^1 S + \frac{i}{g} S^{-1}\partial_\mu S , \tag{20}$$

$$A_\mu^{2\prime} = S^{-1}A_\mu^2 S , \tag{21}$$

$$F_{\mu\nu}^{1\prime} = S^{-1}F_{\mu\nu}^1 S , \tag{22}$$

$$F_{\mu\nu}^{2\prime} = S^{-1}F_{\mu\nu}^2 S , \tag{23}$$

where S is an element in the gauge group G [which is color SU(3) in our case], and A_μ^1 is a matrix in the defining representation of G formed from

$$A_\mu^1 = \underline{A}_\mu^1 \cdot \underline{T} . \tag{24}$$

T_a is a matrix in the defining representation of G satisfying

$$[T_a, T_b] = i f_{abc} T_c , \tag{25}$$

and $\underline{T}$ is a vector formed from such matrices. We note that the homogeneity of the gauge transformation of A_μ^2 allows a mass term to occur in $\mathfrak{L}$ without breaking the gauge symmetry. We shall see that the natural gauge-fixing term to add to the Lagrangian has the form

$$-\frac{1}{\beta}\partial_\mu \underline{A}_\mu^1 \cdot \partial_\nu \underline{A}_\nu^2 . \tag{26}$$

The Euler-Lagrange equations of motion are obtained in the canonical manner:

$$(\partial_\mu + g\underline{A}_\mu^1 \times)\underline{F}_{\mu\nu}^1 - \lambda^2 \underline{A}_\nu^2 = 0 , \tag{27}$$

$$(\partial_\mu + g\underline{A}_\mu^1 \times)\underline{F}_{\mu\nu}^2 + g\underline{A}_\mu^2 \times \underline{F}_{\mu\nu}^1 + g\underline{J}_\nu = 0 , \tag{28}$$

$$\underline{F}_{\mu\nu}^1 = \partial_\mu \underline{A}_\nu^1 - \partial_\nu \underline{A}_\mu^1 + g\underline{A}_\mu^1 \times \underline{A}_\nu^1 , \tag{29}$$

$$\underline{F}_{\mu\nu}^2 = \partial_\mu \underline{A}_\nu^2 - \partial_\nu \underline{A}_\mu^2 + g\underline{A}_\mu^1 \times \underline{A}_\nu^2 - g\underline{A}_\nu^1 \times \underline{A}_\mu^2 , \tag{30}$$

$$(i\nabla + g\not{A}^1 - m)\psi = 0 . \tag{31}$$

The antisymmetry of $\underline{F}_{\mu\nu}^1$ and $\underline{F}_{\mu\nu}^2$ leads to two conservation laws.

$$\partial_\nu(g\underline{A}^1_\mu\times\underline{F}^1_{\mu\nu}-\lambda^2\underline{A}^2_\nu)=0\,, \tag{32}$$

$$\partial_\nu(\underline{A}^1_\mu\times\underline{F}^2_{\mu\nu}+\underline{A}^2_\mu\times\underline{F}^1_{\mu\nu}+\underline{J}_\nu)=0\,, \tag{33}$$

which can be reexpressed as

$$(\partial_\nu+g\underline{A}^1_\nu\times)\underline{A}^2_\nu=0 \tag{34}$$

and

$$(\partial_\nu+g\underline{A}^1_\nu\times)\underline{J}_\nu=0 \tag{35}$$

using the equations of motion. The first of these relations acts in effect as a gauge-fixing term for A^2_μ if a gauge is chosen for A^1_μ. The second relation has the familiar form of current-conservation equations in conventional Yang-Mills theories.

We turn now to the derivation of the perturbation-theory rules in the gluon sector. We consider the vacuum-vacuum transition amplitude in the presence of external sources[9]:

$$W(\underline{J}^1_\mu,\underline{J}^2_\mu)=\int\prod_x d\underline{A}^1_\mu\, d\underline{A}^2_\mu \exp\left[i\int d^4x\left(\mathcal{L}_0-\frac{1}{\beta}\partial_\mu\underline{A}^1_\mu\cdot\partial_\nu\underline{A}^2_\nu+\underline{A}^1_\mu\cdot\underline{J}^1_\mu+\underline{A}^2_\mu\cdot\underline{J}^2_\mu\right)\right]. \tag{36}$$

After some functional translations we find

$$W(\underline{J}^1_\mu,\underline{J}^2_\mu)=\exp\left\{-i\int d^4x\,d^4y\left[\underline{J}^1_\mu(x)\cdot G^{12}_{\mu\nu}(x-y)\cdot\underline{J}^2_\nu(y)+\tfrac{1}{2}\underline{J}^1_\mu(x)\cdot G^{11}_{\mu\nu}(x-y)\cdot\underline{J}^1_\nu(y)\right]\right\}, \tag{37}$$

where we have dropped an irrelevant factor independent of J^1_μ and J^2_μ on the right-hand side, and

$$G^{12}_{\mu\nu ab}(x)=-\delta_{ab}\int\frac{d^4k\,e^{-ik\cdot x}}{(2\pi)^4k^2}\left[g_{\mu\nu}+(\beta-1)\frac{k_\mu k_\nu}{k^2}\right] \tag{38}$$

and

$$G^{11}_{\mu\nu ab}(x)=\frac{\lambda^2\delta_{ab}}{(2\pi)^4}\int\frac{d^4k\,e^{-ik\cdot x}}{k^4}\left[g_{\mu\nu}+(\beta^2-1)\frac{k_\mu k_\nu}{k^2}\right], \tag{39}$$

with a and b labeling color indices. The free propagators corresponding to the time-ordered products are

$$\langle TA^1_{\mu a}(x)A^1_{\nu b}(y)\rangle=iG^{11}_{\mu\nu ab}(x-y) \tag{40}$$

and

$$\langle TA^1_{\mu a}(x)A^2_{\nu b}(y)\rangle=iG^{12}_{\mu\nu ab}(x-y)\,. \tag{41}$$

The somewhat unusual Green's functions of Eqs. (40) and (41) have their origin in the canonical equal-time commutation relations which we shall now find.

From Eqs. (27)–(30) we obtain the equations of motion

$$\partial_0\underline{A}^1_k=\underline{F}^1_{0k}+\partial_k\underline{A}^1_0+g\underline{A}^1_k\times\underline{A}^1_0\,, \tag{42}$$

$$\partial_0\underline{A}^2_k=\underline{F}^2_{0k}+\partial_k\underline{A}^2_0+g\underline{A}^2_k\times\underline{A}^1_0-g\underline{A}^2_0\times\underline{A}^1_k\,, \tag{43}$$

$$\partial_0\underline{F}^1_{0k}=(\partial_i+g\underline{A}_i\times)\underline{F}^1_{ik}-g\underline{A}^1_0\times\underline{F}^1_{0k}+\lambda^2\underline{A}^2_k\,, \tag{44}$$

$$\partial_0\underline{F}^2_{0k}=(\partial_i+g\underline{A}^1_i\times)\underline{F}^2_{ik}-g\underline{A}^1_0\times\underline{F}^2_{0k}-g\underline{A}^2_\mu\times\underline{F}^1_{\mu k}-g\underline{J}_k\,, \tag{45}$$

and equations of constraint

$$\underline{F}^1_{ik}=\partial_i\underline{A}^1_k-\partial_k\underline{A}^1_i+g\underline{A}^1_i\times\underline{A}^1_k\,, \tag{46}$$

$$\underline{F}^2_{ik}=\partial_i\underline{A}^2_k-\partial_k\underline{A}^2_i+g\underline{A}^1_i\times\underline{A}^2_k-g\underline{A}^1_k\times\underline{A}^2_i\,, \tag{47}$$

$$(\partial_i+g\underline{A}^1_i\times)\underline{F}^1_{i0}+\lambda^2\underline{A}^2_0=0\,, \tag{48}$$

$$(\partial_i+g\underline{A}^1_i\times)\underline{F}^2_{i0}+g\underline{A}^2_i\times\underline{F}^1_{i0}-g\underline{J}_0=0\,. \tag{49}$$

The Lagrangian indicates that the canonical momenta are

$$\underline{\Pi}^1_j=\underline{F}^2_{0j} \tag{50}$$

and

$$\underline{\Pi}^2_j=\underline{F}^1_{0j}\,, \tag{51}$$

for $j=1,2,3$ with $\underline{\Pi}^i_j$ conjugate to $\underline{A}^i_j$, and $\underline{A}^i_0$ having no conjugate momentum for $i=1,2$. However, the equations of constraint indicate that not all components are independent. We now find the independent components. Let us define

$$\underline{F}^a_{0i}=\underline{F}^{aT}_{0i}+\underline{F}^{aL}_{0i} \tag{52}$$

and

$$\underline{F}^{aL}_{0i}=\partial_i\underline{\varphi}^a\,, \tag{53}$$

where

$$\partial_i\underline{F}^{aT}_{0i}=0\,. \tag{54}$$

Then Eq. (48) gives

$$(\partial_i+g\underline{A}^1_i\times)\partial_i\underline{\varphi}^1-\lambda^2\underline{A}^2_0=-g\underline{A}^1_i\times\underline{F}^{1T}_{i0} \tag{55}$$

and Eq. (49) gives

$$(\partial_i+g\underline{A}^1_i\times)\partial_i\underline{\varphi}^2+g\underline{A}^2_i\times\partial_i\underline{\varphi}^1 =g\underline{A}^1_i\times\underline{F}^{2T}_{i0}+g\underline{A}^2_i\times\underline{F}^{1T}_{0i}-g\underline{J}_0\,. \tag{56}$$

Rewriting Eqs. (42) and (43) after taking the divergence with respect to spatial components gives

$$(\partial_0+g\underline{A}^1_0\times)\partial_k\underline{A}^1_k=(\partial_k+g\underline{A}^1_k\times)\partial_k\underline{A}^1_0+\partial_k\partial_k\underline{\varphi}^1 \tag{57}$$

and

$$(\partial_0+g\underline{A}^1_0\times)\partial_k\underline{A}^2_k+g\underline{A}^2_0\times\partial_k\underline{A}^1_k =\partial_k\partial_k\underline{A}^2_0+g\underline{A}^2_k\times\partial_k\underline{A}^1_0+g\underline{A}^1_k\times\partial_k\underline{A}^2_0+\partial_k\partial_k\underline{\varphi}^2 \tag{58}$$

If we choose the Coulomb gauge, $\vec{\nabla}\cdot\underline{A}^1=0$, then

$$\partial_k\partial_k\underline{A}_0^1+g\underline{A}_k^1\times\partial_k\underline{A}_0^1+\partial_k\partial_k\underline{\varphi}^1=0 \quad (59)$$

and

$$(\partial_0+g\underline{A}_0^1\times)\partial_k\underline{A}_k^2-\partial_k\partial_k\underline{A}_0^2-g\underline{A}_k^2\times\partial_k\underline{A}_0^1-g\underline{A}_k^1\times\partial_k\underline{A}_0^2 -\partial_k\partial_k\underline{\varphi}^2=0\,, \quad (60)$$

thus determining $\underline{A}_0^1$ and $\underline{A}_0^2$. Suppose we now define

$$\vec{\underline{A}}^2=\vec{\underline{A}}^{2T}+\vec{\underline{A}}^{2L}\,, \quad (61)$$

$$\vec{\underline{A}}^{2L}=\vec{\nabla}\underline{\phi}^3\,, \quad (62)$$

with

$$\vec{\nabla}\cdot\vec{\underline{A}}^{2T}=0 \quad (63)$$

Taking the divergence of Eq. (44) leads to our final equation for dependent variables

$$\lambda^2\partial_k\partial_k\underline{\phi}^3=\partial_0\partial_k\partial_k\underline{\varphi}^1+g\partial_k(\underline{A}_\mu^1\times\underline{F}_{\mu k}^1)\,. \quad (64)$$

The independent dynamical variables are thus seen to be F_{0i}^{1T}, F_{0i}^{2T}, A_i^{1T}, and A_i^{2T}. Their equal-time commutation relations are

$$[F_{0ia}^{1T}(x),A_{jb}^2(y)]=i\delta_{ab}\Delta_{ij}^{tr}(x-y)\,, \quad (65)$$

$$[F_{0ia}^{2T}(x),A_{jb}^1(y)]=i\delta_{ab}\Delta_{ij}^{tr}(x-y)\,, \quad (66)$$

with $i,j=1,2,3$, Δ_{ij}^{tr} given by Eq. (9), and a and b are color indices. All other commutators of the forms $[A^1,A^1]$, $[A^2,A^2]$, $[F^1,F^1]$, $[F^2,F^2]$, $[F^1,F^2]$ are zero.

We return to our development of perturbation-theory rules. The cubic and quartic gluon vertices of our model are given by (see Fig. 3)

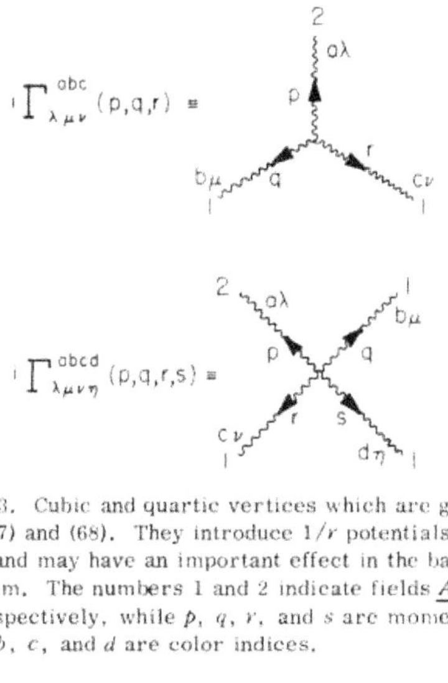

FIG. 3. Cubic and quartic vertices which are given in Eqs. (67) and (68). They introduce $1/r$ potentials in the model and may have an important effect in the baryon spectrum. The numbers 1 and 2 indicate fields $\underline{A}_\mu^1$ and $\underline{A}_\mu^2$, respectively, while p, q, r, and s are momenta, and a, b, c, and d are color indices.

$$i\Gamma^{abc}_{\lambda\mu\nu}(p,q,r)=f^{abc}[g_{\nu\lambda}(r_\mu-p_\mu)+g_{\mu\lambda}(p_\nu-q_\nu) +g_{\mu\nu}(q_\lambda-r_\lambda)]\,, \quad (67)$$

with $p+q+r=0$, and

$$i\Gamma^{abcd}_{\lambda\mu\nu\eta}(p,q,r,s)=-if^{abf}f^{cdf}(g_{\lambda\nu}g_{\nu\eta}-g_{\eta\mu}g_{\lambda\nu}) -if^{acf}f^{bdf}(g_{\lambda\eta}g_{\nu\mu}-g_{\nu\lambda}g_{\mu\eta}) -if^{adf}f^{bcf}(g_{\lambda\eta}g_{\nu\mu}-g_{\lambda\mu}g_{\nu\eta})\,, \quad (68)$$

with $p+q+r+s=0$.

The Faddeev-Popov ghost loops will not be relevant to our line of development so we omit their discussion. The necessity for their introduction[10] is closely related to the requirement of unitarity in Yang-Mills theories. In the present model unitarity will be necessarily violated irrespective of the ghost loops if the Green's functions [Eqs. (38) and (39)] pole ambiguities are resolved by using Feynman's $i\epsilon$ procedure. To avoid unitarity violation we have suggested an alternative procedure where the Green's function singularities are taken in principal value,

$$G^{kL}_{\mu\nu ab}(k^2)=\tfrac{1}{2}[G^{kL}_{\mu\nu ab}(k^2+i\epsilon)+G^{kL}_{\mu\nu ab}(k^2-i\epsilon)]\,, \quad (69)$$

in momentum space (cf. the Appendix). This choice has the advantage stated in the Introduction. The effects are the same as in the Abelian model[3] and may be summarized as: (1) Only states composed solely of quarks contribute to unitarity sums, (2) gluons do not appear in asymptotic states, (3) unitarity is achieved but at the price of possible advanced effects whose range is limited to hadronic dimensions and thus apparently unobservable, and (4) nonscaling corrections to Bjorken scaling in the deep-inelastic electroproduction structure functions are suppressed by a factor of q^2 vis-à-vis the corresponding result using Feynman propagators with q being the virtual photon four-momentum.

A novel feature of the use of principal-value propagators in non-Abelian models is the elimination of closed loops composed solely of gluons. If we consider a subdiagram consisting of a gluon loop with p lines, then Eq. (51) of Ref. 3 gives the Feynman parameter representation

$$I=\int_{-\infty}^{\infty}\prod_{j=1}^{p}\alpha_j\,d\alpha_j\,\frac{\epsilon(\alpha_1\alpha_2\cdots\alpha_p C)}{C^2}\,Ne^{iD/C}\,, \quad (70)$$

where C is a polynomial consisting of Feynman parameters only, while D contains scalar products of external momenta, N symbolizes appropriate numerator factors, and $\epsilon(\alpha)=\pm1$ if $\alpha\gtrless 0$. Since N can be written as a sum of terms each of which is homogeneous in the Feynman parameters, we can take N to be homogeneous without loss of generality. Then scaling all parameters with u, assuming

$$N(u\alpha_1, u\alpha_2, \ldots, u\alpha_p) = u^r N(\alpha_1, \alpha_2, \ldots, \alpha_p), \tag{71}$$

with r an integer, and using

$$\int_0^\infty \frac{du}{u}\,\delta\!\left(1 - \frac{|\alpha_1 + \alpha_2 + \cdots + \alpha_p|}{u}\right) = 1 \tag{72}$$

we find

$$I = \Gamma(r + 2p - 2L) \times \int_{-\infty}^{\infty} \frac{\prod_{j=1}^{p} \alpha_j\, d\alpha_j\, \epsilon(\alpha_1\alpha_2\cdots\alpha_p C) N \delta\left(1 - \left|\sum_k \alpha_k\right|\right)}{C^2(-iD/C)^{r+2p-2L}}, \tag{73}$$

with L = number of loops = 1. Suppose we let $\alpha_j \to -\alpha_j$ for all j in I. Then we find $I = -I$ or

$$I = 0. \tag{74}$$

Thus any closed loop containing only principal-value propagators is zero. Since Faddeev-Popov ghosts appear only in closed loops and consistency[11] requires we use principal-value propagators for them if we use such propagators for gluons, we see that ghosts do not appear in our model. Physically we can understand this result if we remember that ghost loops were introduced to cure problems arising from contributions to unitarity sums of "opened" gluon loops.[10] In our model "opened" loops do not contribute to unitarity sums in any case so the raison d'être for ghosts is lacking.

We now derive the Ward-Takahashi-Slavnov identities using functional methods. Since we take our gluon propagators in principal value it might appear that our use of functional techniques is unjustified. We shall take the view that the functional representation of the vacuum-vacuum transition amplitude embodies the combinatorics of perturbation theory and acts as a generating function for identities, such as the Ward-Takahashi-Slavnov identities. Thus, questions of convergence of functional integrals are irrelevant—the important question is whether identities are valid in perturbation theory.

We define $W(J)$, the vacuum-vacuum transition amplitude, by

$$W(J) = \int \prod_x dA_\mu^1\, dA_\mu^2\, d\psi\, d\bar\psi \exp\left(i \int \tilde{\mathfrak{L}}\, dx\right), \tag{75}$$

with

$$\tilde{\mathfrak{L}} = \mathfrak{L} - \frac{1}{\beta}\partial_\mu \underline{A}_\mu^1 \cdot \partial_\nu \underline{A}_\nu^2 + \underline{A}_\mu^1 \cdot \underline{J}_\mu^1 + \underline{A}_\mu^2 \cdot \underline{J}_\mu^2 + \bar\psi\eta + \bar\eta\psi,$$

with $\mathfrak{L}$ given by Eq. (17). Under the infinitesimal gauge variation

$$\underline{A}_\mu^1 \to \underline{A}_\mu^1 - (\partial_\mu + g\underline{A}_\mu^1 \times)\underline{\theta}, \tag{76}$$

$$\underline{A}_\mu^2 \to \underline{A}_\mu^2 - g\underline{A}_\mu^2 \times \underline{\theta}, \tag{77}$$

$$\psi \to \psi - ig\theta\psi, \tag{78}$$

$$\bar\psi \to \bar\psi + ig\bar\psi\theta, \tag{79}$$

with $\theta = \underline{T}\cdot\underline{\theta}$, $\mathfrak{L}$ is invariant but the remaining terms in $\tilde{\mathfrak{L}}$ lead to

$$\delta\tilde{\mathfrak{L}} = \frac{1}{\beta}[(\partial_\mu + g\underline{A}_\mu^1\times)\partial_\nu\partial_\mu \underline{A}_\nu^2 + g\underline{A}_\nu^2 \times \partial_\mu\partial_\nu \underline{A}_\mu^1]\cdot\underline{\theta} - (\partial_\mu + g\underline{A}_\mu^1\times)\underline{J}_\mu^1\cdot\underline{\theta} - g\underline{J}_\mu^2 \times \underline{A}_\mu^2\cdot\underline{\theta} + ig\bar\psi\theta\eta - ig\bar\eta\theta\psi. \tag{80}$$

Since a transformation of the integration variables does not change the value of the functional integral, the variation of W with respect to θ can be taken to be zero and our equivalent of the Ward-Takahashi-Slavnov identity is

$$\left\{\frac{1}{\beta}\left[D_\nu\left(\frac{\delta}{i\delta \underline{J}_\alpha^1}\right)\partial_\nu\partial_\mu \frac{\delta}{i\delta \underline{J}_\mu^2} + g\frac{\delta}{i\delta \underline{J}_\nu^2} \times \partial_\nu\partial_\mu \frac{\delta}{i\delta \underline{J}_\mu^1}\right] + D_\mu\left(\frac{\delta}{i\delta \underline{J}_\alpha^1}\right)\underline{J}_\mu^1 - g\underline{J}_\mu^2 \times \frac{\delta}{i\delta \underline{J}_\mu^2} + g\underline{T}\eta\frac{\delta}{\delta\eta} - g\bar\eta\underline{T}\frac{\delta}{\delta\bar\eta}\right\} W = 0, \tag{81}$$

with

$$D_\mu\left(\frac{\delta}{i\delta \underline{J}_\alpha^1}\right) = \partial_\mu + g\frac{\delta}{i\delta \underline{J}_\mu^1}\times. \tag{82}$$

In order to investigate the structure of the gluon propagators we shall obtain the proper vertex identity equivalent to Eq. (81). We focus on the novelties of the gluon sector and neglect the quark field terms in $\tilde{\mathfrak{L}}$ and Eq. (81). Let us define

$$W(J) = e^{iZ(J)}, \tag{83}$$

$$\underline{B}_\mu^i = -\frac{\delta Z(J)}{\delta \underline{J}_\mu^i}, \quad i = 1, 2 \tag{84}$$

$$\Gamma(B) = Z(J) + \int d^4x(\underline{J}_\mu^1\cdot \underline{B}_\mu^1 + \underline{J}_\mu^2\cdot\underline{B}_\mu^2), \tag{85}$$

where $\Gamma(B)$ is the generating functional of proper vertices. An immediate consequence is

$$\underline{J}_\mu^i = \frac{\delta\Gamma}{\delta \underline{B}_\mu^i}, \quad i = 1, 2 \tag{86}$$

and as a result Eq. (81) can be rewritten in the form

$$\frac{1}{\beta}\left[\Box\partial_\mu \underline{B}_\mu^2 - g\underline{B}_\nu^1\times\partial_\nu\partial_\mu \underline{B}_\mu^2 - g\underline{B}_\nu^2\times\partial_\mu\partial_\nu \underline{B}_\mu^1 + g\frac{\delta}{i\delta \underline{J}_\nu^1}\times\partial_\nu\partial_\mu \underline{B}_\mu^2 + g\frac{\delta}{i\delta \underline{J}_\nu^2}\times\partial_\nu\partial_\mu \underline{B}_\mu^1\right] - \partial_\mu\frac{\delta\Gamma}{\delta \underline{B}_\mu^1} + \underline{B}_\mu^1\times\frac{\delta\Gamma}{\delta \underline{B}_\mu^1} + \underline{B}_\mu^2\times\frac{\delta\Gamma}{\delta \underline{B}_\mu^2} = 0. \tag{87}$$

If we apply $\delta/\delta \underline{B}_\sigma^1$ to Eq. (87) and set $\underline{B}_\mu^i = 0$ afterwards, we find

$$-\partial_\mu \frac{\delta^2\Gamma}{\delta \underline{B}_\sigma^1 \delta \underline{B}_\mu^1}\bigg|_{B^1=B^2=0} = 0. \tag{88}$$

The second-order functional derivative of Γ is the inverse of the full propagator $G_{\mu\nu ab}^{11\prime}$ and Eq. (88) implies that the proper part of $(G_{\mu\nu ab}^{11\prime})^{-1}$ is purely transverse. We note that the "free" propagator (Eq. 39) contribution to $(G_{\mu\nu ab}^{11\prime})^{-1}$ is not one-particle irreducible and thus not constrained by Eq. (88). Therefore we find the general form

$$G_{\mu\nu ab}^{11\prime}(k) = \delta_{ab}\left(g_{\mu\nu} - \frac{k_\mu k_\nu}{k^2}\right)G^{11}(k^2) + \delta_{ab}\beta^2\lambda^2\frac{k_\mu k_\nu}{k^6}, \tag{89}$$

so that the longitudinal part of the full propagator is not renormalized.

The longitudinal part of the full propagator $G_{\mu\nu ab}^{12\prime}(k)$ is also not renormalized. This may be seen by applying $\delta/\delta \underline{B}_\mu^2$ to Eq. (87) and setting $\underline{B}_\mu^i = 0$ afterwards:

$$\frac{1}{\beta}\Box\partial_\mu\delta^4(x-y) - \partial_\nu\frac{\delta^2\Gamma}{\delta \underline{B}_\mu^2\delta \underline{B}_\nu^1}\bigg|_{B^1=B^2=0} = 0. \tag{90}$$

This implies

$$(G_{\mu\nu ab}^{12\prime})^{-1} = \frac{\delta_{ab}(g_{\mu\nu} - k_\mu k_\nu/k^2)}{G^{12}} - \frac{k_\mu k_\nu\delta_{ab}}{\beta} \tag{91}$$

or

$$G_{\mu\nu ab}^{12\prime}(k) = \delta_{ab}\left(g_{\mu\nu} - \frac{k_\mu k_\nu}{k^2}\right)G^{12}(k) - \beta\delta_{ab}\frac{k_\mu k_\nu}{k^4}. \tag{92}$$

Having now developed the general form of the propagators we now will define the gluon vacuum polarization tensors,

$$\Pi_{\mu\nu ab}^{11}(k) = [G_{\mu\nu ab}^{11\prime}(k)]^{-1} - [G_{\mu\nu ab}^{11}(k)]^{-1}, \tag{93}$$

$$\Pi_{\mu\nu ab}^{12}(k) = [G_{\mu\nu ab}^{12\prime}(k)]^{-1} - [G_{\mu\nu ab}^{12}(k)]^{-1}, \tag{94}$$

which are transverse by our previous discussion:

$$k_\mu\Pi_{\mu\nu ab}^{11} = k_\mu\Pi_{\mu\nu ab}^{12} = 0. \tag{95}$$

Rather than write the Schwinger-Dyson equations for our polarization tensors we have given a diagrammatic representation in Fig. 4.

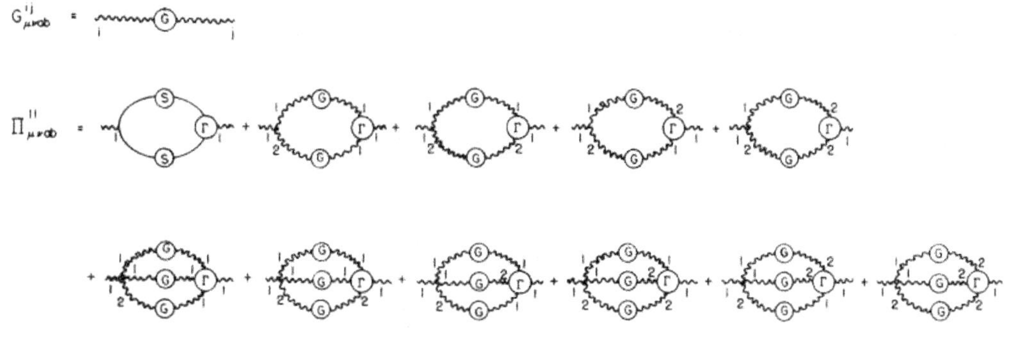

FIG. 4. Diagrammatic representation of the Schwinger-Dyson equation for the proper gluon self-energy, $\Pi_{\mu\nu ab}^{11}$. The numbers at the end of a gluon line specify whether $\underline{A}_\mu^1$ or $\underline{A}_\mu^2$ correspond to that end. The quark propagator is denoted S while Γ denotes the appropriate proper (one-particle irreducible) vertex function. A similar diagrammatic expression can be written for $\Pi_{\mu\nu ab}^{12}$.

IV. OBSERVATIONS

The Schwinger mechanism forces quark confinement to bound color singlet states in a manner which is identical to the Abelian case as described in Sec. II. In order to demonstrate that only color singlets exist in the gauge-invariant physical particle spectrum it is sufficient to show

$$\underline{Q}\psi_{\rm phys}=0\,, \tag{96}$$

where

$$\underline{Q}=\int d^3x\,\underline{J}_0(x) \tag{97}$$

for any physical state $\psi_{\rm phys}$ corresponding to a spatially localized distribution of quarks. We consider a single static quark located at the origin and choose to work in the Coulomb gauge ($\vec{\nabla}\cdot\vec{A}^1=0$). Then the time components of the equations of motion [Eqs. (27) and (28)] lead to (at large distance)

$$\Box\nabla^2\underline{A}_0^1=g\lambda^2\underline{J}_0 \tag{98}$$

if we take into account the elimination of gluons' degrees of freedom through the choice of principal-value propagators and their consequent inability to act as sources. We may now repeat the arguments of Eqs. (13)–(16) for the Abelian case after noticing the occurrence of the Schwinger mechanism in the non-Abelian case which can be verified in low orders of perturbation theory for $\Pi^{11}_{\mu\nu ab}$. Thus the expectation value of the charge in the one-quark state is zero. Since the one-quark state is a charge eigenstate, we find Eq. (96) to be true in this case and more generally through the additivity of the charge operator. Thus only color singlet bound states of quarks are physical.[12]

While the infrared behavior of the theory leads to quark confinement, the ultraviolet behavior allows the quarks to appear quasifree. This is particularly noticeable when we take $\lambda^2=0$ in our Lagrangian and examine the corresponding perturbation theory. Taking $\lambda^2=0$ is equivalent to examining the short-distance behavior of the theory. The only diagrams which exist in this limit are given in Fig. 5. The quark sector of the theory is free. The only nontree structures are one-quark-loop diagrams for the scattering of gluons associated with A_μ^2 (which of course can only be generated by a hypothetical external source). (As a point of comparison we have shown in Fig. 6 the additional diagrams which would occur in the even that Feynman propagators were used—these diagrams necessarily involve gluon loops which principal-value propagators force to be zero.) The vital role of the $\lambda^2A_\mu^2A_\mu^2$ term in the Lagrangian in generating the interacting theory and the fact that λ^2 has the dimensions of $(\text{mass})^2$ allow a natural approximation procedure in this model. This is perhaps best seen within the context of deep-inelastic electroproduction. Just as in the Abelian case we find that the structure functions scale with leading corrections of $O(q^{-4})$, where q equals the virtual photon four-momentum. We can establish a parton picture of scattering wherein the photon is absorbed on one of the quasifree nucleon constituents [as in Fig. 2(a)] if $|q^2|\gg g^2\lambda^2$. Then leading corrections to such a picture [e.g., the diagrams of Fig. 2(b)–2(d)] will be suppressed by $(g^2\lambda^2/q^2)^2$. Thus the dimensional nature of the effective coupling constant allows a particularly simple picture to exist of the region of large spacelike virtual photon mass and the parton picture emerges as a natural approximation.

The k^{-4} form of the quark interaction also appears to have decidedly good features as far as the bound-state structure is concerned. Ignoring the numerator tensor (which does not affect our conclusions), we find the Fourier transform of the gluon propagator,

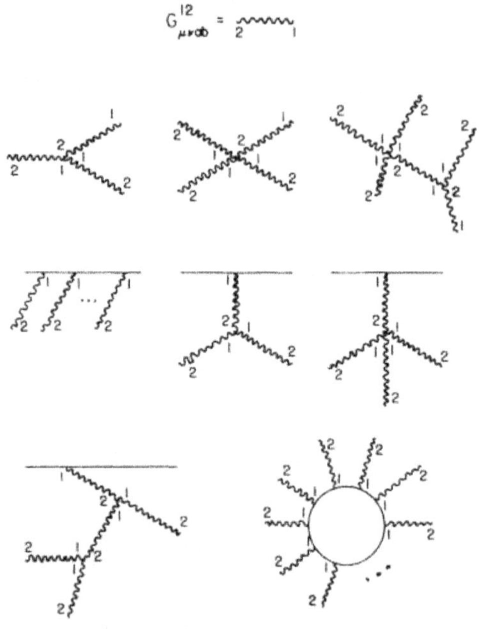

FIG. 5. Some examples of the surviving diagrams in the $\lambda^2=0$ limit of the non-Abelian model with principal-value gluon propagators. Except for the class of one-fermion-loop diagrams only tree diagrams exist in this limit. Note that there are no four (or more) external quark line diagrams and no two (or more) external $\underline{A}_\mu^1$ gluon "external" lines.

$$G(k) = \mathrm{P}\frac{1}{k^4}\,, \qquad (99)$$

to be proportional to

$$\tilde{G}(x) = \theta(x^2)\,. \qquad (100)$$

Since $\tilde{G}$ has a smooth finite limit as $x^2 \to 0$, the short-distance limit, arguments can be made[13] that low-mass bound states can occur in this model. In addition, Dalitz[14] has pointed out that the linearity of trajectories on the Chew-Frautschi plot would follow from a flat-bottomed, smooth interaction—a criterion which is met by Eq. (100). [It is interesting to note that had we used a Feynman propagator rather than principal value, then $\tilde{G}$ would have been $\ln(x^2)$ and thus the general criterion just stated would not have been met. This would appear to be another point in favor of our choice of principal-value propagators.]

Another property which is desirable in the bound-state solutions is nonrelativistic motion of the bound-state constituents.[15] Again an interaction of the form of Eq. (99) appears to realize this feature—even in the strong-binding limit. To see this we shall first take account of the Schwinger mechanism and in the spirit of Hartree-Fock theory modify the quark interaction to

$$G'(k) = \mathrm{P}\frac{1}{(k^2 - \mu^2)^2}\,. \qquad (101)$$

If we now take Eq. (101) to be the Green's function for the effective gluon field and calculate the "Coulomb potential" of a static, point quark source located at the origin we find

$$\varphi(r) = \frac{\varphi_0}{\mu}\, e^{-\mu r}\,, \qquad (102)$$

where ϕ_0 is a constant independent of μ. In the limit $\mu \to 0$ we find

$$\phi(r) \cong \varphi_0\left(\frac{1}{\mu} - r + \cdots\right). \qquad (103)$$

The first two terms of Eq. (103) correspond to choosing Eq. (99) rather than Eq. (101) as the gluon Green's function (in the limit $\mu \to 0$). Equation (102) includes vacuum polarization effects which damp the interaction at large distances. Thus Eq. (102) imperfectly reflects the possibility that a quark-antiquark pair can separate and induce another quark-antiquark pair to be created from the vacuum so that two color singlet mesons will result (presuming it is energetically favored). At shorter distances Eq. (102) appears to be a reasonable approximation. This exponential potential was studied within the framework of the Schrödinger equation in the strong-binding limit (φ_0/μ large) by Greenberg.[16] He showed that the average momentum of the bound constituent in the s state satisfied

$$\frac{p}{m} \sim \left(\frac{\mu}{m}\right)^{1/3} \qquad (104)$$

with m being the quark mass. Thus for μ/m small the quark motion is self-consistently nonrelativistic.

In conclusion, we have shown that a four-dimensional, Lorentz-invariant second-quantized field theory of hadron binding is possible with scaling electroproduction structure functions, only zero-triality physical particle states, and, apparently, linearly rising Regge trajectories and nonrelativistic constituents. A detailed study of the bound states is now in progress.

ACKNOWLEDGMENT

I am grateful to the members of the Newman Laboratory for interesting conversations.

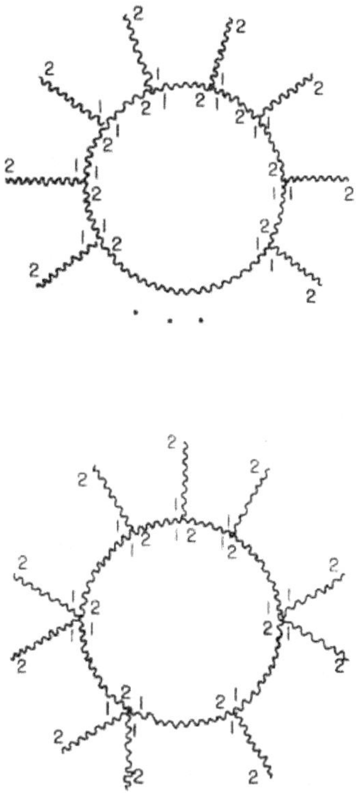

FIG. 6. Some additional diagrams which occur in the $\lambda^2 = 0$ limit of the non-Abelian model if Feynman gluon propagators are used. In addition, there will be Faddeev-Popov ghost-loop diagrams depending on the choice of gauge.

APPENDIX

In Ref. 3 semiclassical arguments based on Dirac's theory of constraints were given to introduce the use of principal-value propagators. We will now describe a second-quantized realization of those arguments for the case of a scalar Klein-Gordon field $\phi(x)$ with the Lagrangian

$$\mathcal{L} = \tfrac{1}{2}(\partial_\mu \varphi)^2 - \tfrac{1}{2}m^2\varphi^2 . \tag{A1}$$

The generalization to vector gluons is immediate. The canonical equal-time commutation relations are

$$[\phi, \varphi] = [\dot{\phi}, \dot{\phi}] = 0 , \tag{A2}$$

$$[\dot{\phi}(\vec{x}, t), \varphi(\vec{y}, t)] = -i\delta^3(\vec{x} - \vec{y}) . \tag{A3}$$

If we expand $\phi(x)$ in plane waves,

$$\varphi(\vec{x}, t) = \sum_{\vec{k}} (A_{\vec{k}} e^{-ik\cdot x} + A^\dagger_{\vec{k}} e^{ik\cdot x}) , \tag{A4}$$

then the q-number Fourier components $A_{\vec{k}}$ must satisfy

$$[A_{\vec{k}}, A_{\vec{k}'}] = [A^\dagger_{\vec{k}}, A^\dagger_{\vec{k}}] = 0 , \tag{A5}$$

$$[A_{\vec{k}}, A^\dagger_{\vec{k}'}] = \delta^3(\vec{k}' - \vec{k}) \tag{A6}$$

for consistency with Eqs. (A2) and (A3). Now the time-ordered product satisfies

$$T(\varphi(x)\varphi(y)) = \epsilon(x_0 - y_0)[\varphi(x), \phi(y)] + \{\phi(x), \phi(y)\} , \tag{A7}$$

with $\epsilon(x_0) = \pm 1$ for $x_0 \gtrless 0$ and $\{A, B\} = AB + BA$. The first term on the right-hand side is a c number completely determined by Eqs. (A5) and (A6). If the second q-number expression were zero, then we would obtain a principal-value propagator from Eq. (A7):

$$T(\varphi(x)\phi(y)) = i\int \frac{d^4k}{(2\pi)^4} e^{-ik\cdot(x-y)} \, \mathrm{P}\, \frac{1}{(k^2 - m^2)} . \tag{A8}$$

We therefore require

$$\{\phi(x), \phi(y)\} = 0 , \tag{A9}$$

with the consequence

$$\{A_{\vec{k}}, A_{\vec{k}'}\} = \{A^\dagger_{\vec{k}}, A^\dagger_{\vec{k}'}\} = \{A_{\vec{k}}, A^\dagger_{\vec{k}'}\} = 0 . \tag{A10}$$

Equations (A5), (A6), and (A10) imply

$$A_{\vec{k}} A_{\vec{k}'} = A^\dagger_{\vec{k}} A^\dagger_{\vec{k}'} = 0 , \tag{A11}$$

$$A_{\vec{k}} A^\dagger_{\vec{k}'} = \tfrac{1}{2}\delta^3(\vec{k} - \vec{k}') , \tag{A12}$$

$$A^\dagger_{\vec{k}} A_{\vec{k}'} = -\tfrac{1}{2}\delta^3(\vec{k} - \vec{k}') \tag{A13}$$

for all $\vec{k}$ and $\vec{k}'$. Thus quadratic terms in A and $A^\dagger$ are reduced to c numbers. It should further be noted that the multiplication rule is not associative.[17] In fact, the multiplication rules of the A and $A^\dagger$ operators in Eqs. (A11)–(A13) are realized by taking multiplication to be

$$UV = \tfrac{1}{2}[U, V] \tag{A14}$$

for U, V being any $A_{\vec{k}}$ or $A^\dagger_{\vec{k}'}$. If we take an analogy to Lie-algebra theory seriously, where the adjoint representation of an algebra has a multiplication rule defined by commutators

$$\tilde{U} * \tilde{V} = [\tilde{U}, \tilde{V}] \tag{A15}$$

then we could call Eqs. (A11)–(A13) the adjoint representation of the Fourier components of φ.

The c-number nature of AA, $A^\dagger A^\dagger$, or $AA^\dagger$ can be understood physically in the following manner. Since the ϕ field has principal-value propagators it is not associated with a particle but is merely the embodiment of an interaction between other objects (which we have suppressed in our Lagrangian). Consequently an emission of a φ field quantum must be directly correlated with a subsequent absorption—it cannot propagate into empty space. The c-number nature of $AA^\dagger$ reflects this correlation between emission and absorption.

Finally, it should be noted that the existence of a vacuum is inconsistent with Eqs. (A11)–(A13).

---

*Work supported in part by the National Science Foundation.

[1]K. Johnson, Phys. Rev. D 6, 1101 (1972); C. M. Bender, J. E. Mandula, and G. S. Guralnik, Phys. Rev. Lett. 32, 1467 (1974); A. Chodos *et al.*, Phys. Rev. D 9, 3471 (1974); W. A. Bardeen *et al.*, *ibid.* 11, 1094 (1975); M. Creutz, *ibid.* 10, 1749 (1974); P. Vinciarelli, Nuovo Cimento Lett. 4, 905 (1972); R. Dashen, B. Hasslacher, and A. Neveu, Phys. Rev. D 10, 4114 (1974); 10, 4130 (1974); 10, 4138 (1974).

[2]Y. Nambu, in ***Preludes in Theoretical Physics***, edited by A. de-Shalit, H. Feshbach, and L. Van Hove (North-Holland, Amsterdam, 1966), p. 133; H. J. Lipkin, Phys. Lett. 45B, 267 (1973).

[3]S. Blaha, Phys. Rev. D 10, 4268 (1974).

[4]J. Schwinger, Phys. Rev. 128, 2425 (1962).

[5]A. Pais and G. Uhlenbeck, Phys. Rev. 79, 145 (1950); J. Kiskis, Phys. Rev. D 11, 2178 (1975).

[6]A. Casher, J. Kogut, and L. Susskind, Phys. Rev. D 10, 732 (1974); J. Lowenstein and J. Swieca, Ann. Phys. (N.Y.) 68, 172 (1971).

[7]R. Jackiw and G. Preparata, Phys. Rev. Lett. 22, 975 (1969); S. Adler and W. Tung, *ibid.* 22, 978 (1969); S. Blaha, Phys. Rev. D 3, 510 (1971).

[8]The Lagrangian of Eq. (17) was first written by D. Sinclair as a generalization of the Abelian model of Ref. 3. An alternative non-Abelian model for quark confinement has been suggested by S. K. Kauffmann [Nucl. Phys. B87, 133 (1975)]. I am grateful to Dr. Kauffmann for sending me a copy of his paper prior to publication.

[9]E. Abers and B. W. Lee [Phys. Rep. 9C, 1 (1973)] provide a useful review of conventional Yang-Mills theories.

[10]R. P. Feynman, Acta Phys. Pol. 24, 697 (1963).

[11]B. W. Lee and J. Zinn-Justin [Phys. Rev. D 5, 3121 (1972)] point out that the $i\epsilon$ prescription in their Eq. (2.8) for the ghost loop is dictated by unitarity considerations.

[12]This does not preclude color-singlet states of the gluons from playing a role in the theory. They are not particles but can be exchanged between color-singlet quark states in scattering events. On naive dimensional grounds they should be most important in forward scattering. This leads to the possibility that the Pomeron might possibly be interpreted as a "two-gluon bound state". In the case of wide-angle scattering the predominant mechanism for large momentum transfer would appear to be constituent interchange due to the strong damping effects of k^{-4} propagators on momentum transfer.

[13]M. Böhm, H. Joos, and M. Krammer, in *Recent Developments in Mathematical Physics*, proceedings of the XII Schladming Conference (Acta Phys. Austriaca Suppl. XI), edited by P. Urban (Springer, New York, 1970), p. 3.

[14]R. H. Dalitz, a paper presented at the Topical Conference on Meson Spectroscopy, Philadelphia, 1968 (unpublished).

[15]H. J. Lipkin, Phys. Rep. 8C, 175 (1973).

[16]O. W. Greenberg, Phys. Rev. 147, 1077 (1966).

[17]Nonassociative field operators have been previously used by M. Günaydin and F. Gürsey, Phys. Rev. D 9, 3387 (1974).

9. Spin 1 Higgs Mechanism

Higgs fields are usually treated as spin 0 fields. However it is possible for higher spin fields to have non-zero vacuum expectation values. We saw a non-Abelian example in the preceding chapter. In this chapter we develop a Higgs Mechanism formulation for spin 1 fields using our pseudoquantization method.

9.1 Pseudoquantization of Spin 1 Fields

We begin by defining two neutral vector fields that correspond to the scalar particle: $\varphi_{1\mu}(x)$ and $\varphi_{2\mu}(x)$. These fields will be assumed to have the equal time commutators

$$[\varphi_{i\mu}(x), \pi_{j\nu}(y)] = ig_{\mu\nu}(1 - \delta_{ij})\delta^3(\mathbf{x} - \mathbf{y}) \tag{9.1}$$
$$[\varphi_{i\mu}(x), \varphi_{j\nu}(y)] = 0$$
$$[\pi_{i\mu}(x), \pi_{j\nu}(y)] = 0$$

where δ_{ij} is the Kronecker δ and where $\pi_{i\mu}(x)$ is the canonically conjugate momentum to $\varphi_{i\mu}(x)$. The fields $\varphi_{1\mu}(x)$ and $\pi_{1\mu}(y)$ will be observable classical fields similar to those in eqs. 69 and 70 in Appendix 2-A. Appendix 2-A provides a more detailed and comprehensive discussion of pseudoquantization as used in this chapter. The fields $\varphi_{2\mu}(x)$ and $\pi_{2\mu}(y)$ will not be observables. Thus $\varphi_{1\mu}(x)$ and $\pi_{1\mu}(y)$ can both be sharp on the set of physical states.

We now specify the lagrangian density for a generic spin 1 particle:

$$\mathcal{L} = \partial\varphi_{1\nu}/\partial x_\mu \partial\varphi_2{}^\nu/\partial x^\mu \tag{9.2a}$$

with hamiltonian density

$$\mathcal{H} = \pi_{1\nu}\, \pi_2{}^\nu + \partial\varphi_{1\nu}/\partial x_i \partial\varphi_2{}^\nu/\partial x^i \tag{9.2b}$$

where i labels spatial coordinates, and $\pi_{1\nu} = \partial\varphi_{2\nu}/\partial t$ and $\pi_{2\nu} = \partial\varphi_{1\nu}/\partial t$. Eqs. 9.2 are without a potential or mass term.

The lagrangian and hamiltonian for a massive boson are

$$\mathcal{L} = \partial\varphi_{1\nu}/\partial x_\mu \partial\varphi_2{}^\nu/\partial x^\mu - m^2\, \varphi_{1\nu}\varphi_2{}^\nu \tag{9.2c}$$

with hamiltonian density

$$\mathcal{H} = \pi_{1\nu}\pi_2{}^\nu + \partial\varphi_{1\nu}/\partial x_i \partial\varphi_2{}^\nu/\partial x^i + m^2\, \varphi_{1\nu}\varphi_2{}^\nu \tag{9.2d}$$

The fields can be fourier expanded in terms of creation and annihilation operators:

$$\varphi_{i\nu}(\mathbf{x}, t) = \int d^3k\, [a_{i\nu}(k)f_k(x) + a_{i\nu}{}^\dagger(k)f_k{}^*(x)] \tag{9.3}$$

for i = 1, 2 where

$$f_k(x) = e^{-ik\cdot x}/(2\omega_k(2\pi)^3)^{1/2}$$

with $\omega_k = (|\mathbf{k}|^2 + m^2)^{1/2\,i}$ for a massive vector boson.

The creation and annihilation operators satisfy the commutation relations:

$$[a_{i\mu}(k), a_{j\nu}^{\dagger}(k')] = (1 - \delta_{ij})g_{\mu\nu}\delta^3(\mathbf{k} - \mathbf{k}') \tag{9.4}$$
$$[a_{i\mu}(k), a_{j\nu}(k')] = 0$$
$$[a_{i\mu}^{\dagger}(k), a_{j\nu}^{\dagger}(k')] = 0$$

for i, j = 1, 2. The vacuum state |0> satisfies

$$a_{1\nu}(k)|0> = a_{1\nu}^{\dagger}(k)|0> = 0 \tag{9.5}$$
$$a_{2\nu}(k)|0> \neq 0 \qquad\qquad a_{2\nu}^{\dagger}(k)|0> \neq 0 \tag{9.6}$$

for all ν. The dual vacuum state satisfies

$$<0|a_{2\nu}(k) = <0|a_{2\nu}^{\dagger}(k) = 0 \tag{9.7}$$
$$<0|a_{1\nu}(k) \neq 0 \qquad\qquad <0|a_{1\nu}^{\dagger}(k) \neq 0 \tag{9.8}$$

for ν = 0, 1, 2, 3. Positive energy single particle *ket* states are defined using $a_{2\nu}^{\dagger}(k)$ while negative energy ket states are defined using $a_{2\nu}(k)$. Positive energy single particle *bra* states are defined using $a_{1\nu}(k)$ while negative energy bra states are defined using $a_{1\nu}^{\dagger}(k)$.

9.2 Classical Field States for Spin 1 Bosons

The defining properties of a classical vector field state are:

$$\varphi_{1\nu}(x)|\Phi, \Pi> = \Phi_\nu(x)|\Phi, \Pi> \tag{9.9}$$
$$\pi_{1\nu}(x)|\Phi, \Pi> = \Pi_\nu(x)|\Phi, \Pi>$$

where $\Phi_\nu(x)$ and $\Pi_\nu(x)$ are sharp on the states and where $\varphi_{1\nu}(x)$ is given by eq. 9.3. A classical c-number vector field has the form

$$\Phi_\nu(\mathbf{x}, t) = \int d^3k\, [\alpha_\nu(k)f_k(x) + \alpha_\nu^*(k)f_k^*(x)] \tag{9.10}$$

The corresponding classical state is a coherent state with the form

$$|\,\Phi, \Pi> = C\exp\left\{\int d^3k\, [\alpha_\nu(k)a_2^{\nu\dagger}(k) + \alpha_\nu^*(k)a_2^{\nu}(k)]\right\}|0> \tag{9.11}$$

and correspondingly for $\Pi_\nu(x)$ where C is a normalization constant.

In order to obtain a constant vacuum expectation value in the rest frame of a vector particle we define the "vacuum" state:

$$|\Phi, \Pi> = C\exp\{[(2\pi)^3 m/2]^{1/2}\Phi_v[a_2^{v\dagger}(\mathbf{0},m) + a_2^{v}(\mathbf{0},m)]\}|0> \quad (9.11a)$$

Φ_v is a constant 4-vector in the Higgs particle rest frame. *Thus our pseudoquantum formalism allows us to define coherent vector "vacuum" states that can lead to particle masses and interaction term constants.*

Additional details on coherent states, which differ from conventional coherent states such as those of Kibble[77] and others, can be found in Appendix 2-A.

9.3 Spin 1 Higgs Mechanism

With the pseudoquantum coherent state formalism, which gives purely classical fields, and also quantum fields through the use of φ_{2v} and its creation and annihilation operators, we now have the machinery to define a mass mechanism without the introduction of a potential whose origin can only be described as dubious.

For we can define a pseudoquantum coherent state that yields a constant, non-zero vacuum expectation value:

$$\varphi_{1v}(x)|\Phi, \Pi> = \Phi_v|\Phi, \Pi> \quad (9.12)$$

where Φ_v is a constant for all v. Evaluating part of a fermion interaction term we find

$$\bar{\psi}(\varphi_{1v} + \varphi_{2v})\psi \rightarrow \bar{\psi}(\Phi_v + \varphi_{2v})\psi \quad (9.13)$$

It can also generate a mass through an interaction with a gauge field of the form

$$A^{\mu}(\varphi_{1v} + \varphi_{2v})^2 A_{\mu} \rightarrow A^{\mu}(\Phi_v + \varphi_{2v})^2 A_{\mu} \quad (9.14)$$

The present formalism[78] *thus provides a clean way to separate the vacuum expectation value of a vector particle from its quantum vector field part.*

---

[77] T. W. B. Kibble, Jour. Math. Phys. **2**, 212 (1961).

[78] To obtain both the vacuum expectation value and the interaction with the quantum part of the pseudoquantum fields we choose to always specify interactions with fermions and gauge fields using both fields together: $\varphi_v = \varphi_{1v} + \varphi_{2v}$ as seen above.

REFERENCES

Bjorken, J. D., Drell, S. D., 1964, *Relativistic Quantum Mechanics* (McGraw-Hill, New York, 1965).

Bjorken, J. D., Drell, S. D., 1965, *Relativistic Quantum Fields* (McGraw-Hill, New York, 1965).

Blaha, S., 1998, *Cosmos and Consciousness* (Pingree-Hill Publishing, Auburn, NH, 1998).

_______, 2002, *A Finite Unified Quantum Field Theory of the Elementary Particle Standard Model and Quantum Gravity Based on New Quantum Dimensions™ & a New Paradigm in the Calculus of Variations* (Pingree-Hill Publishing, Auburn, NH, 2002).

_______, 2003, *A Finite Unified Quantum Field Theory of the Elementary Particle Standard Model and Quantum Gravity Based on New Quantum Dimensions™ and a New Paradigm in the Calculus of Variations* (Pingree-Hill Publishing, Auburn, NH, 2003).

_______, 2004, *Quantum Big Bang Cosmology: Complex Space-time General Relativity, Quantum Coordinates™ Dodecahedral Universe, Inflation, and New Spin 0, ½, 1 & 2 Tachyons & Imagyons* (Pingree-Hill Publishing, Auburn, NH, 2004).

______, *2005a, Quantum Theory of the Third Kind: A New Type of Divergence-free Quantum Field Theory Supporting a Unified Standard Model of Elementary Particles and Quantum Gravity based on a New Method in the Calculus of Variations* (Pingree-Hill Publishing, Auburn, NH, 2005).

______, 2005b, *The Metatheory of Physics Theories, and the Theory of Everything as a Quantum Computer Language* (Pingree-Hill Publishing, Auburn, NH, 2005).

______, 2005c, *The Equivalence of Elementary Particle Theories and Computer Languages: Quantum Computers, Turing Machines, Standard Model, Superstring Theory, and a Proof that Gödel's Theorem Implies Nature Must Be Quantum* (Pingree-Hill Publishing, Auburn, NH, 2005).

______, 2006a, *The Foundation of the Forces of Nature* (Pingree-Hill Publishing, Auburn, NH, 2006).

______, 2006b, *A Derivation of ElectroWeak Theory based on an Extension of Special Relativity; Black Hole Tachyons; & Tachyons of Any Spin.* (Pingree-Hill Publishing, Auburn, NH, 2006).

______, 2007a, *Physics Beyond the Light Barrier: The Source of Parity Violation, Tachyons, and A Derivation of Standard Model Features* (Pingree-Hill Publishing, Auburn, NH, 2007).

______, 2007b, *The Origin of the Standard Model: The Genesis of Four Quark and Lepton Species, Parity Violation, the ElectroWeak Sector, Color SU(3), Three Visible Generations of Fermions, and One Generation of Dark Matter with Dark Energy* (Pingree-Hill Publishing, Auburn, NH, 2007).

______, *2008a, A Direct Derivation of the Form of the Standard Model From GL(16) (Pingree-Hill Publishing, Auburn, NH, 2008).*

______, 2008b, *A Complete Derivation of the Form of the Standard Model With a New Method to Generate Particle Masses Second Edition* (Pingree-Hill Publishing, Auburn, NH, 2008)

______, 2009, *The Algebra of Thought & Reality: The Mathematical Basis for Plato's Theory of Ideas, and Reality Extended to Include A Priori Observers and Space-Time Second Edition* (Pingree-Hill Publishing, Auburn, NH, 2009).

______, 2010a, *Operator Metaphysics: A New Metaphysics Based on a New Operator Logic and a New Quantum Operator Logic that Lead to a Mathematical Basis for Plato's Theory of Ideas and Reality* (Pingree-Hill Publishing, Auburn, NH, 2010).

______, 2010b, *The Standard Model's Form Derived from Operator Logic, Superluminal Transformations and GL(16)* (Pingree-Hill Publishing, Auburn, NH, 2010).

______, 2011a, *21st Century Natural Philosophy Of Ultimate Physical Reality* (McMann-Fisher Publishing, Auburn, NH, 2011).

______, 2011b, *All the Universe! Faster Than Light Tachyon Quark Starships & Particle Accelerators with the LHC as a Prototype Starship Drive Scientific Edition* (Pingree-Hill Publishing, Auburn, NH, 2011).

______, 2011c, *From Asynchronous Logic to The Standard Model to Superflight to the Stars* (Blaha Research, Auburn, NH, 2011).

______, 2012a, *From Asynchronous Logic to The Standard Model to Superflight to the Stars volume 2: Superluminal CP and CPT, U(4) Complex General Relativity and The Standard Model, Complex Vierbein General Relativity, Kinetic Theory, Thermodynamics* (Blaha Research, Auburn, NH, 2012).

______, 2012b, *Standard Model Symmetries, And Four And Sixteen Dimension Complex Relativity; The Origin Of Higgs Mass Terms* (Blaha Reasearch, Auburn, NH, 2012).

______, 2013a, *Multi-Stage Space Guns, Micro-Pulse Nuclear Rockets, and Faster-Than-Light Quark-Gluon Ion Drive Starships* (Blaha Research, Auburn, NH, 2013).

______, 2013b, *The Bridge to Dark Matter; A New Sister Universe; Dark Energy; Inflatons; Quantum Big Bang; Superluminal Physics; An Extended Standard Model Based on Geometry* (Blaha Reasearch, Auburn, NH, 2013).

______, 2014a, *Universes and Multiverses: From a New Standard Model to a Physical Multiverse; The Big Bang; Our Sister Universe's Wormhole; Origin of the Cosmological Constant, Spatial Asymmetry of the Universe, and its Web of Galaxies; A Baryonic Field between Universes and Particles; Flatverse Extended Wheeler-DeWitt Equation* (Blaha Reasearch, Auburn, NH, 2014).

______, 2014b, *All the Multiverse! Starships Exploring the Endless Universes of the Cosmos Using the Baryonic Force* (Blaha Research, Auburn, NH, 2014).

______, 2014c, *All the Multiverse! II Between Multiverse Universes: Quantum Entanglement Explained by the Multiverse Coherent Baryonic Radiation Devices – PHASERs Neutron Star Multiverse Slingshot Dynamics Spiritual and UFO Events, and the Multiverse Microscopic Entry into the Multiverse* (Blaha Research, Auburn, NH, 2014).

______, 2015a, *PHYSICS IS LOGIC PAINTED ON THE VOID: Origin of Bare Masses and The Standard Model in Logic, U(4) Origin of the Generations, Normal and Dark Baryonic Forces, Dark Matter, Dark Energy, The Big Bang, Complex General Relativity, A Megaverse of Universe Particles* (Blaha Research, Auburn, NH, 2015).

______, 2015b, *PHYSICS IS LOGIC Part II: The Theory of Everything, The Megaverse Theory of Everything, U(4)⊗U(4) Grand Unified Theory (GUT), Inertial Mass = Gravitational Mass, Unified Extended Standard Model and a New Complex General Relativity with Higgs Particles, Generation Group Higgs Particles* (Blaha Research, Auburn, NH, 2015).

______, 2015c, *The Origin of Higgs ("God") Particles and the Higgs Mechanism: Physics is Logic III, Beyond Higgs – A Revamped Theory With a Local Arrow of Time, The Theory of Everything Enhanced, Why Inertial Frames are Special, Universes of the Mind* (Blaha Research, Auburn, NH, 2015).

______, 2015d, *The Origin of the Eight Coupling Constants of The Theory of Everything: U(8) Grand Unified Theory of Everything (GUTE), S^8 Coupling Constant Symmetry, Space-Time*

Dependent Coupling Constants, Big Bang Vacuum Coupling Constants, Physics is Logic IV (Blaha Research, Auburn, NH, 2015).

______, 2016a, *New Types of Dark Matter, Big Bang Equipartition, and A New U(4) Symmetry in the Theory of Everything: Equipartition Principle for Fermions, Matter is 83.33% Dark, Penetrating the Veil of the Big Bang, Explicit QFT Quark Confinement and Charmonium, Physics is Logic V* (Blaha Research, Auburn, NH, 2016).

______, 2016b, *The Periodic Table of the 192 Quarks and Leptons in The Theory of Everything: The U(4) Layer Group, Physics is Logic VI* (Blaha Research, Auburn, NH, 2016).

Eddington, A. S., 1952, *The Mathematical Theory of Relativity* (Cambridge University Press, Cambridge, U.K., 1952).

Fant, Karl M., 2005, *Logically Determined Design: Clockless System Design With NULL Convention Logic* (John Wiley and Sons, Hoboken, NJ, 2005).

Heitler, W., 1954, *The Quantum Theory of Radiation* (Claendon Press, Oxford, UK, 1954).

Huang, Kerson, 1992, *Quarks, Leptons & Gauge Fields 2nd Edition* (World Scientific Publishing Company, Singapore, 1992).

Misner, C. W., Thorne, K. S., and Wheeler, J. A., 1973, *Gravitation* (W. H. Freeman, New York, 1973).

Sagan, H., 1993, *Introduction to the Calculus of Variations* (Dover Publications, Mineola, NY, 1993).

Sakurai, J. J., 1964, *Invariance Principles and Elementary Particles* (Princeton University Press, Princeton, NJ, 1964).

Streater, R. F. and Wightman, A. S., 2000, *PCT, Spin, Statistics, and All That* (Princeton University Press, Princeton, NJ 2000).

Weinberg, S., 1972, *Gravitation and Cosmology* (John Wiley and Sons, New York, 1972).

Weinberg, S., 1995, *The Quantum Theory of Fields Volume I* (Cambridge University Press, New York, 1995).

Weyl, H., 1950, *Space, Time, Matter* (Dover, New York, 1950).

Weyl, H., (Tr. S. Pollard et al), 1987, *The Continuum* (Dover Publications, New York, 1987).

INDEX

About the Author

Stephen Blaha is a well-known Physicist and Man of Letters with interests in Science, Society and civilization, the Arts, and Technology. He had an Alfred P. Sloan Foundation scholarship in college. He received his Ph.D. in Physics from Rockefeller University. He was a faculty member at a number of US universities: an Instructor at Univ. Wash. (Seattle) and Cornell, a faculty member at Syracuse, Williams, and Yale, and for 20 years an Associate of the Harvard Physics Dept. employed at Harvard in the period 1983-2003.

He was also a Member of the Technical Staff at Bell Laboratories, a manager at the Boston Globe Newspaper, a Director at Wang Laboratories, and President of Blaha Software Inc and of Janus Associates Inc. (NH).

Among other achievements he was a co-discoverer of the "r potential" for heavy quark binding developing the first (and still the only demonstrable) non-abelian gauge theory with an "r" potential; first suggested the existence of topological structures in superfluid He-3; first proposed Yang-Mills theories would appear in condensed matter phenomena with non-scalar order parameters; first developed a grammar-based formalism for quantum computers and applied it to elementary particle theories; first developed a new form of quantum field theory without divergences (thus solving a major 60 year old problem that enabled a unified theory of the Standard Model and Quantum Gravity without divergences to be developed); first developed a formulation of complex General Relativity based on analytic continuation from real space-time; first developed a generalized non-homogeneous Robertson-Walker metric that enabled a quantum theory of the Big Bang to be developed without singularities at $t = 0$; first generalized Cauchy's theorem and Gauss' theorem to complex, curved multi-dimensional spaces; received Honorable Mention in the Gravity Research Foundation Essay Competition in 1978; first developed a physically acceptable theory of faster-than-light particles; first derived a composition of extrema method in the Calculus of Variations; first quantitatively suggested that inflationary periods in the history of the universe were not needed; first proved Gödel's Theorem implies Nature must be quantum; provided a new alternative to the Higgs Mechanism, and Higgs particles, to generate masses; first showed how to resolve logical paradoxes including Gödel's Undecidability Theorem by developing Operator Logic and Quantum Operator Logic; first developed a quantitative harmonic oscillator-like model of the life cycle, and interactions, of civilizations; first showed how equations describing superorganisms also apply to civilizations. A recent book shows his theory applies successfully to the past 14 years of history and to *new* archaeological data on Andean and Mayan civilizations as well as Early Anatolian and Egyptian civilizations.

He first developed an axiomatic derivation of the forms of The Standard Model from geometry – space-time properties – The Extended Standard Model. It has a Dark Matter sector that approximates the ElectroWeak sector with Dark doublets and Dark gauge interactions. It also uses quantum coordinates to remove infinities that crop up in most interacting quantum field theories and additionally to remove the infinities that appear in the Big Bang and generate an inflationary growth of the universe. The Extended Standard Model has an ultra-high energy GUT (Grand Unified Theory) limit with a $U(4)\otimes U(4)$ symmetry; and can be united with gravitation to form a Theory of Everything. (See *Physics is Logic Part II.*)

Blaha has had a major impact on a succession of elementary particle theories: his Ph.D. thesis (1970), and papers, showed that quantum field theory calculations to all orders in ladder approximations could not give scaling deep inelastic electron-nucleon scattering. He later showed the eigenvalue equation for the fine structure constant α in Johnson-Baker-Willey QED had a zero at $\alpha = 1$ not 1/137 by solving the Schwinger-Dyson equations to all orders in an approximation that agreed with exact results to 4th order in α thus ending interest in this theory. In 1979 at Prof. Ken Johnson's (MIT) suggestion he calculated the proton-neutron mass difference in the MIT bag model and found the result had the wrong sign reducing interest in the bag model. These results all appear in Physical Review papers. In the 2000's he repeatedly pointed out the shortcomings of SuperString theory and showed that The Standard Model's form could be derived from space-time geometry by an extension of Lorentz transformations to faster than light transformations. This deeper space-time basis greatly increases the possibility that it is part of THE fundamental theory.

In graduate school (1965-71) he wrote substantial papers in elementary particles and group theory: The Inelastic E- P Structure Functions in a Gluon Model. Phys. Lett. B40:501-502,1972; Deep-Inelastic E-P Structure Functions In A Ladder Model With Spin 1/2 Nucleons, Phys.Rev. D3:510-523,1971; Continuum Contributions To The Pion Radius, Phys. Rev. 178:2167-2169,1969; Character Analysis of U(N) and SU(N), J. Math. Phys. 10, 2156 (1969); and The Calculation of the Irreducible Characters of the Symmetric Group in Terms of the Compound Characters, (Published as Blaha's Lemma in D. E. Knuth's book: *The Art of Computer Programming Vols. 1 – 4*).

In the early 1980's Blaha was also a pioneer in the development of UNIX for financial, scientific and Internet applications: benchmarked UNIX versions showing that block size was critical for UNIX performance, developing financial modeling software, starting database benchmarking comparison studies, developing Internet-like UNIX networking (1982) and developing a hybrid shell programming technique (1982) that was a precursor to the PERL programming language. He was also the manager of the AT&T ten-year future products development database. His work helped lead to commercial UNIX on computers such as Sun Micros, IBM AIX minis, and Apple computers.

In the 1980's he pioneered the development of PC Desktop Publishing on laser printers. and was nominated for three "Awards for Technical Excellence" in 1987 by PC Magazine for PC software products that he designed and developed.

Recently he has developed a theory of Megaverses – actual universes of which our universe is one – with quantum particle-like properties based on the Wheeler-DeWitt equation of Quantum Gravity. He has developed a theory of a baryonic force, which had been conjectured many years ago, and estimated the strength of the force based on discrepancies in measurements of the gravitational constant G. This force, operative in 15-dimensinal space, can be used to escape from our universe in "uniships" which are the equivalent of the faster-than-light starships proposed in the author's earlier books. Thus travel to other universes, as well as to other stars is possible.

Blaha also considered the complexified Wheeler-DeWitt equation and showed that its limitation to real-valued coordinates and metrics generated a Cosmological Constant in the Einstein equations.

The author has also recently written a series of books on the serious problems of the United States and their solution as well as a book on the decline of Mankind that will follow from current social and genetic trends in Mankind.

In the past twelve years Dr. Blaha has written over 40 books on a wide range of topics. Some recent major works are: *From Asynchronous Logic to The Standard Model to Superflight to the Stars*, *All the Universe!*, *SuperCivilizations: Civilizations as Superorganisms*, *America's Future: an Islamic Surge, ISIS, al Qaeda, World Epidemics, Ukraine, Russia-China Pact, US Leadership Crisis*, *The Rises and Falls of Man – Destiny – 3000 AD: New Support for a Superorganism MACRO-THEORY of CIVILIZATIONS From CURRENT WORLD TRENDS and NEW Peruvian, Pre-Mayan, Mayan, Anatolian, and Early Egyptian Data, with a Projection to 3000 AD*, and *Mankind in Decline: Genetic Disasters, Human-Animal Hybrids, Overpopulation, Pollution, Global Warming, Food and Water Shortages, Desertification, Poverty, Rising Violence, Genocide, Epidemics, Wars, Leadership Failure.*

In recent books in the *Physics is Logic Series* he derived an Extended Standard Model from Asynchronous Logic in complex space-time, showed the structure of the known Standard Model naturally follows, showed Dark Matter, Dark Energy, the Big Bang and inflation naturally appear in the enlarged theory, expanded the model into a Theory of Everything in which inertial mass was shown equal to gravitational mass and mass contributions from gravity appear in elementary fermions, Considering Higgs particles and bosons in general he developed a theory called pseudoquantization that naturally yielded Higgs-like symmetry breaking, a local arrow of time, a reason for the special significance of inertial reference frames, and a mechanism for making coupling constants into boson fields vacuum expectation values that are possibly time-dependent. Further he showed that another U(4) symmetry exists that makes our known fermion spectrum with four generations into four duplicate layers (three of which are unknown at present but constitute part of Dark Matter) that are connected by an as yet undiscovered ultra-weak interaction. Assuming an Equipartition Principle for fermions at the Big Bang time the theory yields 83.33% as the Dark Matter percentage of the total matter in the universe – approximately the same as astrophysical experiments. Lastly he showed the phenomenological Charmonium potential follows directly from his 1974 theory of the Strong Interaction, and showed the Stong coupling is only a factor of three times the electromagnetic fine structure constant making Strong interaction perturbation theory feasible.

He has taught approximately 4,000 students in undergraduate, graduate, and postgraduate corporate education courses primarily in major universities, and large companies and government agencies.

The above paragraphs summarize much of his work over the past fifty years. This work is fully documented. He continues to engage in research and writing at Blaha Research.

www.ingramcontent.com/pod-product-compliance
Lightning Source LLC
Chambersburg PA
CBHW082009190326
41458CB00010B/3124
* 9 7 8 0 9 9 7 0 7 6 1 3 4 *